BIOLOGY'S BEGINNINGS

This edition first published in Great Britain 2023 by Walker Books Ltd,
87 Vauxhall Walk, London SE11 5HJ in collaboration with The MIT Press

2 4 6 8 10 9 7 5 3 1

Cover images: Courtesy of Biblioteca Apostolica Vaticana, Città del Vaticano (front centre); Courtesy of the National Library of Medicine (back centre); front and back: The Walters Art Museum (top left), Library of Congress, Rare Book & Special Collections Division (top right and bottom left), Vue du Cajambe, Wellcome Collection, Attribution 4.0 International (CC BY 4.0) (bottom right)

Additional image credits appear on pages 179-181

This book was typeset in Alegreya and Candara

Printed in China

British Library Cataloguing in Publication Data:
a catalogue record for this book is available from the British Library

ISBN 978-1-5295-1221-2

MITeen Press, an imprint of Walker Books Ltd

miteenpress.com
walker.co.uk

DISCOVERING LIFE'S STORY

VOLUME ONE

BIOLOGY'S BEGINNINGS

JOY HAKIM

MiTeenPress

Contents

al bahr al šāmi uahūa al rūmi
bahr al hozar
ard al kumania
ard al rusia
ard bulġār min al turk

INTRODUCTION

The Islamic World, the Christian World, and Some Foundational Life Science

In *this* beginning, it is the eighth century, and things are happening in the new city of Baghdad, which will eventually become part of modern Iraq. The city straddles the Tigris River and is home to the Caliph al-Mansur, a Muslim ruler with big ideas: he intends to make Baghdad the center of a great empire.

What is the caliph like? According to Karen Armstrong, a Catholic theologian known for her incisive books on religion, "Courtiers kissed the ground when they came into his presence in a way that would have been unimaginable in the days when Arabs prostrated themselves only before God."

LEFT: A map from 1154, drawn by Muhammad al-Idrisi, a geographer and cartographer who was born in Ceuta, a Spanish city in North Africa. The map, which is upside down from a modern perspective, was commissioned by King Roger II of Sicily.

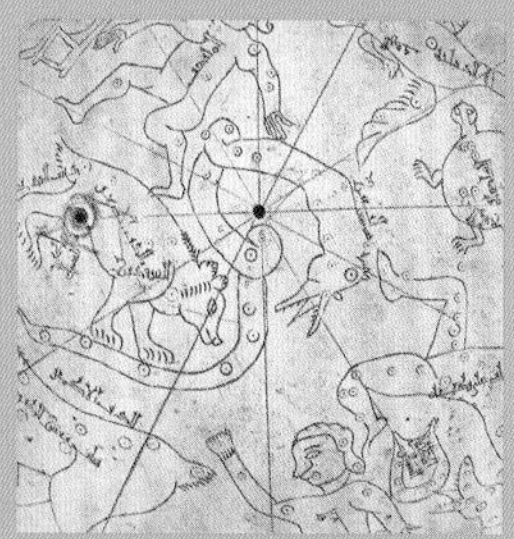

This celestial globe, built in Valencia, Spain, in 1085 by Ibrahim ibn Said al-Sahli, is the oldest known celestial globe. Today it is in the Museo Galileo, in Florence, Italy. The lower image is a close-up view.

The circular ancient city of Baghdad as imagined by a modern artist, Jean Soutif

Yakut al-Hamawi (1179–1229), a formerly enslaved Arab writer, describes Baghdad as

> *two vast semicircles on the right and left banks of the Tigris, twelve miles in diameter. The numerous suburbs, covered with parks, gardens, villas, and beautiful promenades, and plentifully supplied with rich bazaars, and finely built mosques and baths, stretched for a considerable distance on both sides of the river.*

The Muslim empire is expanding across the Middle East and North Africa and on to Spain and Portugal. This will be known as Islam's Golden Age. It will last from the eighth to the thirteenth century, bringing centers of learning to Cairo, Egypt; and Córdoba, Spain; as well as to Baghdad.

About four thousand years earlier, the ancient city of Uruk, some 150 miles (240 kilometers) south of Baghdad, archived written materials, including the *Epic of Gilgamesh*,

This Stone Age carving shows Gilgamesh and Enkidu slaying the monster Humbaba, which was considered a historic event. The Epic of Gilgamesh *is one of the earliest stories of friendship.*

the heart-wrenching story of one of Uruk's kings, in a temple library. Two very old cities, Nineveh and Babylon, also have renowned libraries. None hold paper books; rather, these libraries house inscribed clay tablets. One advantage of clay: if a library burns, some of the books will survive. In Egypt, in the city of Alexandria, a library founded in the third century BCE becomes known through much of the Eurasian world; writings there are kept on rolls of papyrus.

Papyrus is a plant that grows on watery land along the Nile River. Strips of papyrus are laid in two layers, one horizontal and one vertical, and then pressed and dried to form a paper-like sheet.

Al-Hamawi, who spends ten years traveling in Persia, Syria, and Egypt, leaves notes on what he sees. Perhaps because he is a reader, he focuses on libraries and bookstores.

Egyptian artisans harvesting and splitting papyrus, which, when processed, will serve as writing paper. This painting is based on a scene in the tomb of the noble Puimre at Thebes.

This region, which will be known later as part of the Middle East, has a long tradition of supporting libraries. The best of the libraries are like universities: they attract brilliant minds. In Baghdad, thinkers are said to congregate at a fine library called the House of Wisdom. When the House of Wisdom is destroyed during the 1258 Mongol siege of Baghdad, many of its books are said to have been thrown into the Tigris River. Later, its role as a center of learning will be debated. Still, there is no doubt that ancient writings that might otherwise have been lost are preserved in the Muslim world's libraries.

The library in Alexandria, Egypt, held one of the most important collections of knowledge in the ancient world. Much of its collection was burned in a siege of the city by Julius Caesar. This is a scene of the library as imagined by the artist J. André Castaigne, who drew it in 1908.

What is a control group? In a scientific experiment, a control group does not receive experimental treatment. Then its participants can be compared with those who did receive treatment, in order to judge the treatment's effectiveness.

This book centers on life science in the Western world. But knowledge doesn't pay attention to boundaries—on land or in books. Scholars in the Islamic Golden Age ask questions about life science and medicine that help establish those fields. They preserve and study the writings of ancient Greece's pioneering philosopher, Aristotle, and Rome's remarkable Galen, a Greek-speaking physician. They add their own scholarship: Muhammad ibn Zakariya al-Razi, a doctor, differentiates smallpox from measles; he also pioneers the use of a control group in experiments.

Abu Ali Sina, known in Europe as Avicenna, lives from 980 to 1037. He is one of the first doctors in the West credited with monitoring the heart's rhythms by squeezing the wrist to feel a person's pulse. A philosopher, astronomer, and poet as well as a doctor, he compiles a vast

Aristotle and Plato: Different Ways to Consider Life Science

For the great Greek thinker Plato (427–347 BCE), people are imitations of an idealized, perfect human form, even if this perfect form is never visible to us. The same theory applies, in Plato's view, to all animals and objects. A dog has its own idealized form; so does a banana.

Plato's brilliant student, Aristotle (384–322 BCE), rejects Plato's idea. He closely observes life around him, particularly in a lagoon on the island of Lesbos, in the Aegean Sea. There he studies the differences between creatures such as dolphins and fish. He watches as a chicken embryo develops. His approach, blending theory with observation, lays foundations for what will be called the scientific method.

Knowledge develops in fits and starts, and not always in tidy steps. The writings of both Plato and Aristotle are translated from ancient Greek and studied by writers in the Islamic world and then eventually become popular again in the European world.

Plato inspires students to be creative and imaginative.

Aristotle becomes a father of life science.

medical encyclopedia called the *Canon of Medicine*; it draws on the medical traditions of China and India as well as those from his own milieu and will be used for centuries in Europe (especially in Spain) and in the Islamic world.

Things are different to the west. The great art, literature, and science of the Greek and Roman golden era, which began in the fifth century BCE, is hardly remembered. In Europe, the pursuit of the arts and higher learning has largely fallen by the wayside. To the north, seafaring Vikings are terrorizing settlements in the British Isles and on Europe's Atlantic coast.

In the eleventh century, when a university opens in Bologna, it is the first institution of its kind in Europe. Other universities in other cities soon follow. The universities are hungry for ideas to teach. Writings by Christians and Muslims soon flow back and forth from Islamic lands in the Middle East to Europe's universities. Scholars in

Islamic Spain and Sicily participate in an open society where writers and thinkers from Muslim, Christian, and Jewish backgrounds contribute ideas and share knowledge.

Thomas Aquinas, depicted in a fifteenth-century altarpiece by Carlo Crivelli, originally situated in Ascoli Piceno, Italy

At the new University of Paris, Thomas Aquinas, a Christian scholar who lives from 1225 to 1274, attempts to bring together his Christian faith and Aristotle's ideas on science. More broadly, an entwining of Greco-Roman arts with Christian philosophy, Islamic scholarship, and the Hebrew Bible will provide a creative foundation for much that is to come in the Western world. In what may seem to us to be a technology-deprived era, ideas move slowly, but they do move.

In the 1330s, an Italian poet, Francesco Petrarch (1304–1374), does something unusual. He climbs a mountain—Mount Ventoux, in southern France—just to look at the scenery. Petrarch notes what he sees and then attempts to use that knowledge to draw conclusions. Is this the beginning of a new kind of thinking in medieval Europe? Some say yes.

Petrarch has been born into Europe's Middle Ages, a time when only a few people can read and plagues are regular, lethal horrors. The sixth-century plague of Justinian, for example, named for the Byzantine emperor of Constantinople, is believed to have killed tens of millions of people from Europe to Asia to Africa, which was much of the world's population at the time.

How do people deal with such mass death? Many turn to religion in hopes that it can save them—or offer an explanation. Most of those who look to Christianity to explain their world can't actually read the Bible. What is known as the Bible's Old Testament is written in Hebrew, and the Bible's New Testament is in Latin. There are no other translations, so hardly anyone understands what they are hearing when it is read aloud in church.

In the fourteenth century, an Oxford University professor named John Wycliffe begins to translate the Bible into English so that ordinary English people, not just the priests, can know what it says. But Wycliffe's work is not authorized by the king or the established church (which in England at that time is Catholic). When an earthquake

This 1497 painting by the artist Josse Lieferinxe shows Saint Sebastian pleading to Jesus for the lives of the afflicted during a seventh-century outbreak of the bubonic plague.

occurs off the English coast, some believe God is commenting on Wycliffe's translation. Wycliffe's writings are banned in England, and more than four decades after his death, his bones are dug up and burned.

In 1450, a German inventor named Johannes Gutenberg invents a printing press, which changes life for those who can read and for many who will learn to read. Before the printing press, books had to be hand printed, which made them very expensive. Moderately priced books will soon help produce a new generation of readers. Before long, coffeehouses open in most of Europe's cities, and people gather in them to talk about books and politics. After all the plagues and all the deaths, many are ready to ask big questions and consider big ideas. And some want to read the Bible.

In 1525, another Oxford scholar, William Tyndale, who also believes people should be able to read the Bible for themselves, goes to Germany to publish his English translation

This mural by John W. Alexander is part of a series on the evolution of books at the Library of Congress, in Washington, DC. Here, a worker produces printed pages, one of which is being examined by the printer Johannes Gutenberg.

of the Old and New Testaments. The idea of a Bible that can be widely read remains deeply controversial, and early copies have to be smuggled into England. Tyndale is later burned at the stake for heresy. But change is coming. In 1522, a German priest, Martin Luther, translates the New Testament into German. And then, in 1611, King James of England himself sponsors an English-language Bible. Now the printing presses across Europe stay busy turning out copies of the Bible that have been translated in languages Europeans actually speak. And that makes many want to learn to read, which they do.

A page from a Bible printed by Johannes Gutenberg, circa 1454

Europe's new readers have a whole lot to learn about. They can read of the Venetian merchant Marco Polo's thirteenth-century journey across Asia's silk route to China, where he was received by the great Yuan (or Mongol) ruler Kublai Khan. What Polo saw and put into his book makes some of his contemporaries study maps; before this, most have known very little about the world beyond their neighborhoods.

THE
HOLY
BIBLE,
Conteyning the Old Testament,
AND THE NEW:

ANNO DOM. 1611

From the 1611 edition of the King James Bible

For centuries, Europeans do not know that there are thriving nations in what is now called the Americas, or even that those continents exist. The Aztecs are building pyramids, the Incas are building thousands of miles of roads along South America's western coast, and many Indigenous nations—including the Powhatan, the Mohican, the Algonquin, the Lakota, and many more—are flourishing in North America.

Beginning in 1492 with the voyage of Christopher Columbus, Europeans learn of lands previously unknown to them; they will name the region America after a mapmaker who drew what he thought were its boundaries. The Europeans send shiploads of people who carry firearms to the Americas: their goal is to take possession of gold and silver, and eventually the land itself, an invasion that will decimate the Indigenous peoples of the Americas. Along with guns, the invaders bring diseases that will kill staggering numbers of Indigenous people, who have never encountered the germs Europeans carry and have no immunity to them. For Indigenous peoples, the devastation caused by European microbes is one catastrophic effect of contact with Europeans.

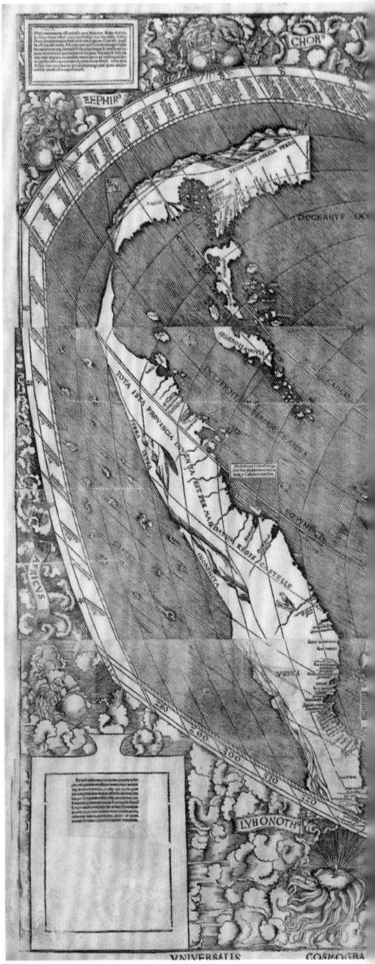

We don't know how many Indigenous people in the Americas die of disease after contact with Europeans. Some estimates say upward of fifty million, which is 90 percent of the population. The depopulation reshapes the very environment, including cultivated areas, irrigation systems, forests carefully managed by controlled fires, and huge stone structures. Despite this devastation, which will take centuries to unfold, Indigenous cultures, nations, languages, and people will find ways to resist and continue—and their ongoing vitality and presence will be very much evident on these continents.

But there is far more than just conquest in this era. Perhaps it is the plague deaths, perhaps it is the new information about the Americas, perhaps it is the moderately priced books; whatever the reason, some Europeans open their eyes and question old ideas. That questioning becomes central in the emerging Renaissance, a time between the fourteenth and seventeenth centuries when classical philosophy and art that had once flourished in ancient Greece and Rome are

This 1507 map by the German cartographer Martin Waldseemüller is the first to show a distinct Pacific Ocean and features American lands only recently visited by European explorers.

rediscovered, and painting, sculpture, poetry, music, and literature flourish. So too does the study of science.

Innovative concepts from emerging science now combine with artistry to challenge ordinary thinking, producing a period of mind-jolting newness. In Italy, a world-class genius named Leonardo da Vinci (1452–1519) is born into what has become a vibrant

Renaissance. Leonardo is a painter, sculptor, inventor, and self-motivated medical scholar, perhaps unique in world history.

At a time when medical school dissections are mostly of dogs and oxen, Leonardo performs at least thirty human dissections—many at night, by candlelight. Intending to document the human body, he does some 750 precise anatomical drawings: of bone structures, muscles, organs, and the brain. One of his exquisite drawings is of a human fetus inside a womb. Wanting to understand the human eye, Leonardo hard-boils one in an egg white, which allows him to dissect it with extraordinary precision.

Leonardo da Vinci's sketch of a fetus in its mother's womb is a groundbreaking medical rendering as well as gorgeous art. Leonardo got the position of the fetus in the uterus right, which no one is known to have done in a drawing before.

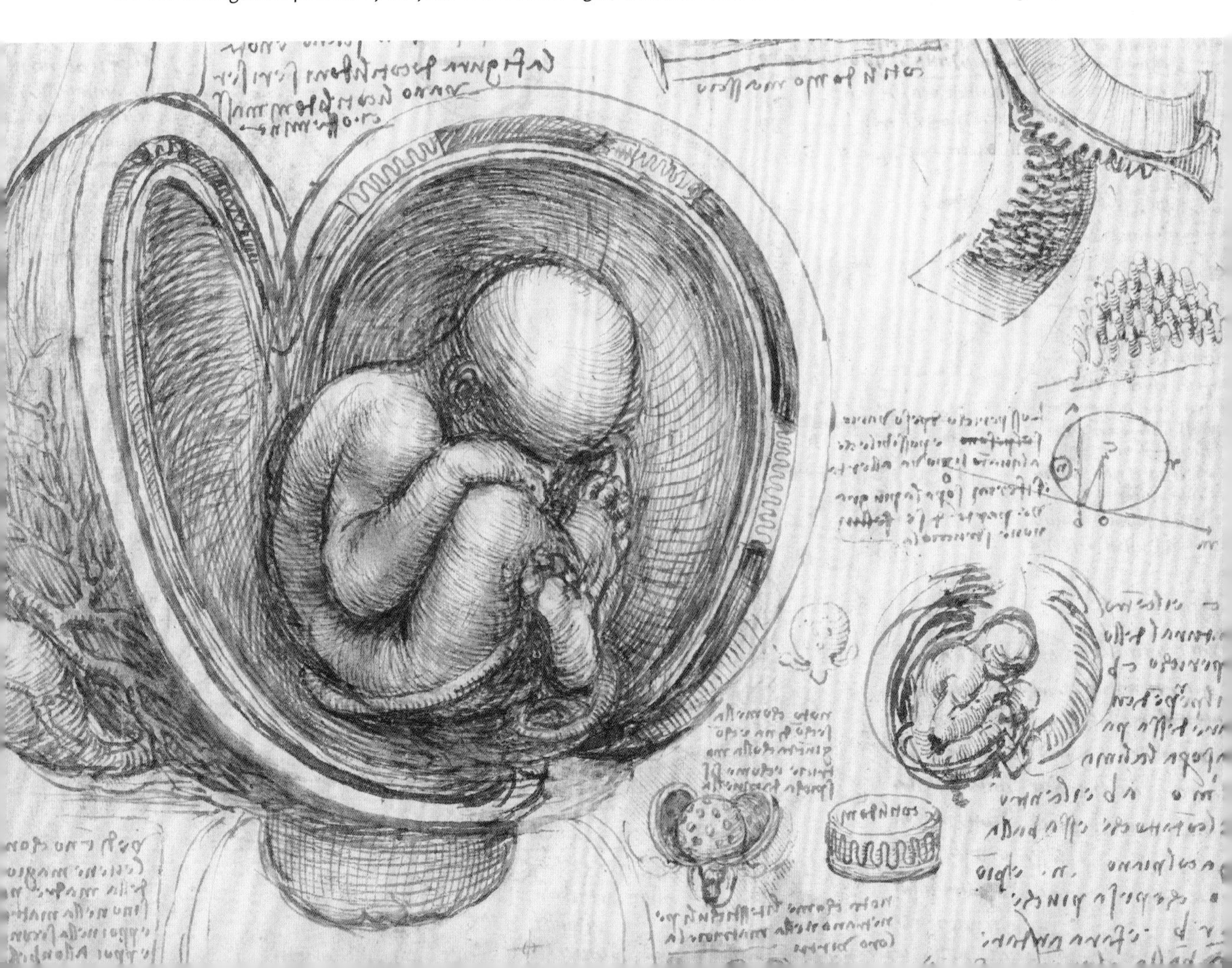

Up until that time, Europeans believed, as the revered Roman physician Galen had, that blood originates in the liver, and that the heart is the source of the body's heat and soul. "Leonardo was among the first to fully appreciate that the heart, not the liver, was the center of the blood system," writes his biographer Walter Isaacson. "Leonardo was able to show, contrary to Galen, that the heart is simply a muscle rather than some form of special vital tissue. Like all muscles, the heart has its own blood supply and nerves."

His discoveries are not universally welcomed. When Pope Leo X learns about da Vinci's drawings, he orders him to stop. But Leonardo has opened a door that can't be closed.

We have learned much about life since that time. We now know that all life-forms grow and develop by following a code with instructions embedded in the cells of every living thing—this means that life follows a master plan with clear rules. Those rules are followed by your cells and mine, as well as by the cells of turnips and tigers.

No one in the Renaissance understands that. Unlocking the secrets of life's code will take time, and it won't be easy. But serious inquiries into life science during the Renaissance will jump-start the process. And so will ideas that come from tinkerers: people creating new devices and ways of thinking that will lead to technological breakthroughs.

It is after the Renaissance, beginning in 1665, that magnifying microscopes first appear, enabling humans to glimpse life's actual building blocks, known as cells. Over time, microscopes, which use glass lenses to enlarge images, will get better and better. Still, it won't be until the far-off twentieth century, when a new kind of electron microscope is invented, that we will be able to peer inside the components that make up cells.

Amazing things are going to happen; astounding information is going to be discovered. But not yet.

Right now, it is the 1530s, we are in France, and an independent-minded medical student is preparing to challenge his professor—and centuries of medical tradition. ■

A Scientific Superstar, a Plague, and an Influential Artist-Thinker

01

The rise of the West is unparalleled in world history. The countries north of the Alps had for centuries been regarded as a backward region, which had attached itself to Greco-Roman culture of the south and had, gradually, developed its own distinctive version of Christianity and its own form of agrarian culture. Western Europe lagged far behind the Christian empire of Byzantium, where the Roman Empire had not collapsed as it had in Europe.

—Karen Armstrong

The first new science grounded in what we would call facts was the anatomy of Vesalius (1543), which depended on the public space of the anatomy theatre and the public space of the printed book to rebut the previously uncontested authority of Galen.

—David Wootton

LEFT: In this drawing, Vesalius has cut open the stomach of a woman who has just been hanged. Many in the crowd of observers are there to learn; some are there to pick pockets. Vesalius will hold up body parts and explain their function. No one here knows anything about germs or the importance of cleanliness. And, except for the victim and a nun seen praying, everyone here is male.

At the University of Paris, Johann Guenther (1505–1574), a German physician known in his home country as Johann Winter von Andernach, sits in a thronelike chair reading aloud from a masterwork on the anatomy of the human body written by the great Greek physician Galen. Medical education in Europe is built on Galen's writings, and Guenther's translation into Latin from Greek is said to be the best there is. The students who are listening to the professor intend to be doctors.

An Important Place of Learning

The University of Paris, also known as the Sorbonne, is named for Robert de Sorbon (1201–1274), a priest who was King Louis IX's confessor. It soon draws scholars and students from across Europe, becoming a vibrant center of thinking. Some say it is because of the university that Paris is chosen as the capital of France.

It is 1581, and the English surgeon Dr. John Banister (1533–1610) is giving a lecture in London's Barber-Surgeons' Hall.

Why a barber? Barbers need to be careful where they trim, shave, and cut or they will lose customers! Barbers are therefore among those most experienced at handling knives.

It is 1534, and thanks to Gutenberg and his printing press, invented almost a hundred years before, books are widely available in Europe, but not so widely that every student can have his own copy of books like *Principles of Anatomy according to the Opinion of Galen*.

As the professor reads, a nearby surgeon cuts into a dead animal. An instructor uses a stick to point out body parts. Neither professor nor instructor gets blood on his hands. It is the surgeon, often a barber, who is bloodied. Once a year, the dissection is of a human, but the rest of the year, it may be an ox, a sheep, or a dog that gets sliced.

A twenty-year-old medical student, Andreas Vesalius (1514–1564), is among Guenther's students. Vesalius has had a privileged childhood. His father is royal apothecary to the Holy Roman Emperor Charles V. His British mother is an educated woman from a family of physicians.

Vesalius speaks, writes, and thinks in Latin, Greek, and Hebrew, as well as in French and English. He also thinks for himself, and he is not impressed with the esteemed professor or the way he is teaching anatomy.

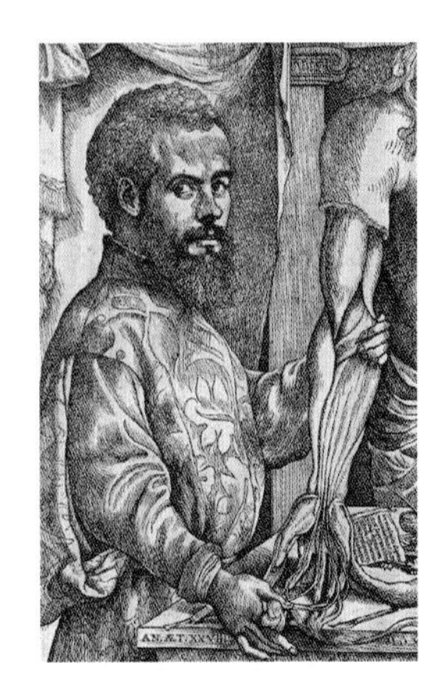
A portrait of Vesalius

Vesalius has been dissecting mice and small animals since he was a kid; he has extraordinary skills when it comes to using a knife, and he wants to dissect human cadavers. As it happens, Paris is filled with dead bodies—city officials string up thieves and other criminals on street corners.

So he scrounges cadavers and does his own dissections, in an era when human dissections are often controversial and sometimes banned. When some teachers and students hear what he is doing, they ask him to cut into a human corpse and explain its parts to them, which he does. At first this does not please the renowned Johann Guenther, especially when Vesalius finds errors in the revered Galen's work.

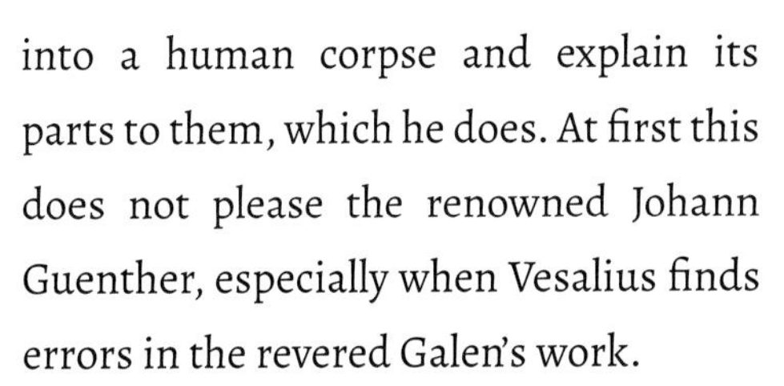

During the Renaissance, tradespeople, merchants, and craftsmen receive the formal schooling necessary for their work (generally about what we would consider fifth-grade level today). Professions requiring a university education, such as medicine and the law, are mostly filled by a male elite. Education, or a lack of it, keeps many people from professions where they might do well.

In 1536, war breaks out between France and the Holy Roman Empire; Vesalius is forced to head home to Belgium. There, at the University of Leuven, he does what is said to be the first human dissection performed in Belgium in eighteen years. But he leaves Belgium for the University of Padua, which is a good move for a medical student. Italy, at the center of one of history's great inventive moments, is generating creative minds—both academic and artistic. The medical school there is said to be as good as they get.

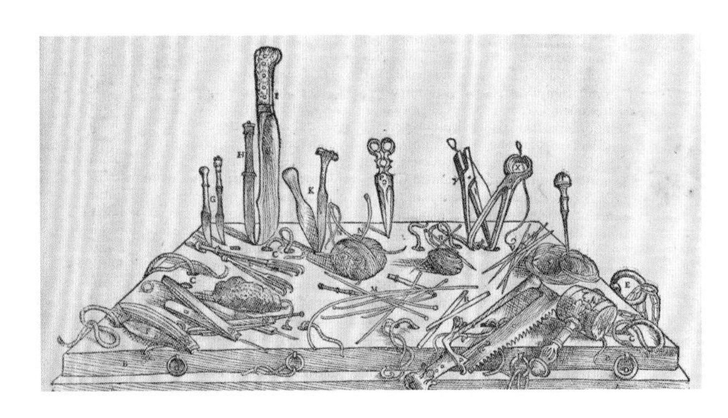
A drawing of surgical instruments from Vesalius's great work, the Fabrica

Who Was Galen?

This portrait of the ancient Greek physician Galen is from a sixteenth-century woodcut.

Galen was part of a group of Greek and Roman thinkers whose work was studied by Muslim scholars in the Middle Ages and shared with European academics. Originally from Greece, Galen lived from 129 to 216 CE and became the Roman Empire's most renowned physician; he was doctor to Rome's emperors and to its gladiators.

Galen's wealthy parents made sure that their son was well educated; he was able to travel the Mediterranean lands and study ways to treat illness. He had an innovative mind, and he paid attention to what he observed. He saw a relationship between diet and health. He was also aware of the importance of cleanliness when few other people were. He became the author of works on anatomy and on medical practice that would provide a foundation for much that was to come in the world of doctoring. But in ancient Rome, human dissections were forbidden, so as far as we know, Galen never dissected a human cadaver. Mostly he dissected oxen and monkeys, which is why he got a few things wrong about human anatomy.

Some thirteen centuries after Galen lived, Guenther reads aloud from his work in teaching students of medicine. And, as Vesalius realizes, some of what he teaches them is wrong.

In Padua, Vesalius continues his human dissections, often before large audiences. The day he receives his doctorate in medicine, he is asked to stay at Padua as chair of surgery and anatomy. A judge will provide him with bodies of executed criminals. Sometimes executions are timed to coordinate with his lectures, since refrigeration does not yet exist.

Vesalius has dead bodies hung from a beam in the lecture hall. As he cuts into the corpse (he does it himself, not using a barber), he points out body parts. His classes are jammed with practicing physicians as well as university students.

What his students see often contradicts the long-dead Galen's words. Vesalius writes, "Galen was deluded by his dissection of ox brains and described the cerebral vessels, not of a human but of oxen." And in the same volume, he states, "Aristotle in particular, and quite a few others, thought that the nerves took origin from the heart." Finally, Vesalius makes the point that "Galen never inspected a human uterus."

Vesalius hires an artist to do eleven large, detailed teaching charts drawn from human cadavers he has dismembered. The identity of the artist is not entirely clear, but the drawings are works of art as well as illustrations

The University of Leuven will celebrate its six hundredth birthday in 2025. It is one of the oldest universities in the world and the oldest Catholic university. Desiderius Erasmus was a professor there in the fifteenth century. In 1497, he says this about teaching in a letter to his friend Christian Northoff:

> *A constant element of enjoyment must be mingled with our studies, so that we think of learning as a game rather than a form of drudgery, for no activity can be continued for long if it does not to some extent afford pleasure to the participant.*

When I undertake the dissection of a human cadaver I pass a stout rope . . . beneath the lower jaw and through the two zygomas [bony arches on the side of the head] up to the top of the head. . . . The longer end . . . I run through a pulley fixed to a beam in the room so that I may raise or lower the cadaver as it hangs there or may turn it round in any direction to suit my purpose.

—Andreas Vesalius

Vesalius's importance to medicine still looms large. Here he is on a 1989 postage stamp from Hungary.

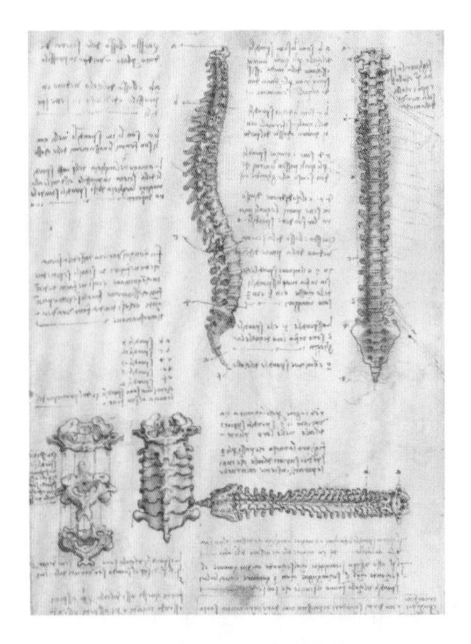

Leonardo da Vinci's innovative drawings of a human spine informed doctors and non-doctors alike; they were part of his detailed study of human anatomy.

intended to teach anatomy. They become landmarks in art as well as life science and education history.

Vesalius expects to reach a still bigger audience with a new book he plans to publish.

He may have been inspired by an unpublished project that was part of a collaboration between Leonardo da Vinci and Marcantonio della Torre (1481–1511), a professor of anatomy at the University of Padua. They worked together on what was to be a masterwork detailing the human body. Della Torre provided some twenty human cadavers, which his students had dissected. Leonardo used them to draw illustrations for the anatomy book they planned. Leonardo's biographer Walter Isaacson says he "made 240 drawings and wrote at least thirteen thousand words of text, illustrating and describing every bone, muscle group, and major organ in the human body." Today we have some, but not all, of Leonardo's drawings.

His drawings include the first accurate depiction of the human spine. He does drawings of animal body parts to contrast with the anatomy of humans. But in 1511, a pandemic known as the plague sweeps through Italy and much of Europe, and della Torre dies. Leonardo briefly finds safety in the Vatican in Rome, where the artists Michelangelo and Raphael are at work creating masterpieces while they also attempt to escape the horrible plague. No one knows its cause or how to deal with it. But they do understand that it is passed from person to person.

Leonardo da Vinci's story continues in 1516, when Francis I of France invites him to live in a château in Amboise, next

Plagues are major historic events in Europe, with three horrendous outbreaks from the 1300s to the 1800s. What becomes known as the Black Death is a bubonic plague that, at its height from 1347 to 1351, is said to have killed half the population of Europe. (One of its symptoms is swelling in the lymph nodes in the armpit or groin.) Since no one knows about bacteria, many believe that God is punishing Europeans for sinful behavior.

Plagues and epidemics have peppered human history. This is a depiction of plague victims in ancient Roman times.

A second wave of bubonic plague appears in the 1500s. And in the seventeenth century—between 1629 and 1631—as many as one in every four Italians dies of bubonic plague.

In England, in the seventeenth century, Cambridge University sends its students home when the plague appears. One of those students, Isaac Newton, uses his time at home to think about the heavens and Earth and develop ideas later found in his great book the *Principia*, about how the universe works. Classical mechanics, also known as Newtonian physics, begins with that book.

to the royal residence. Then and now, Amboise is a river town with sunshine, fresh breezes, and dramatic vistas. Leonardo heads there, taking some of his anatomical drawings with him.

In Amboise, Leonardo is able to continue his scientific research. "The spinal cord is the source of the nerves that gives voluntary movement to the limbs," he writes. And: "The muscles which move the lips are more numerous in man than in any other animal." He figures out (long before most doctors) that it is the heart that pumps blood around the body, not the liver. He understands and describes blood flow centuries before it is accepted medical knowledge. In fleeing the pandemic, he has found a patron, Francis I, who, awed by Leonardo's talents and intellect, supports his work. Here he

The Château du Clos Lucé, Leonardo da Vinci's home in Amboise

has the freedom to do what he wants with his time and mind. But not for long.

Leonardo dies in Amboise in 1519.

What becomes of his notes and his drawings of the human body? Most disappear until the nineteenth century, when someone checking the Royal Collection in London finds six hundred drawings by Leonardo da Vinci. Today those drawings are celebrated for their artistic mastery.

Does Vesalius see any of Leonardo's drawings? We don't know. He almost certainly is aware of Leonardo's intent to detail the human body.

Leonardo has been dead for twenty-four years when, in 1543, Vesalius publishes a seven-volume masterpiece: *De Humani Corporis Fabrica Libri Septum* (*On the Structure of the Human Body*). Often known simply as the *Fabrica*, this illustrated study of the human body combines anatomical insights with stunning artistic renderings.

There is some debate about who did the famous drawings in Vesalius's landmark book. We believe they were done at the workshop, known as an atelier (pronounced ah-tel-YAY), of the great artist Titian. They may be the work of Jan Stephan van Calcar, a student of Titian's, or of someone else in Titian's studio, or perhaps even of Vesalius himself—the experts don't agree on this. They do agree that Van Calcar carved the life drawings into the giant woodblocks that Vesalius uses to create posters that teach anatomy to students in his classroom.

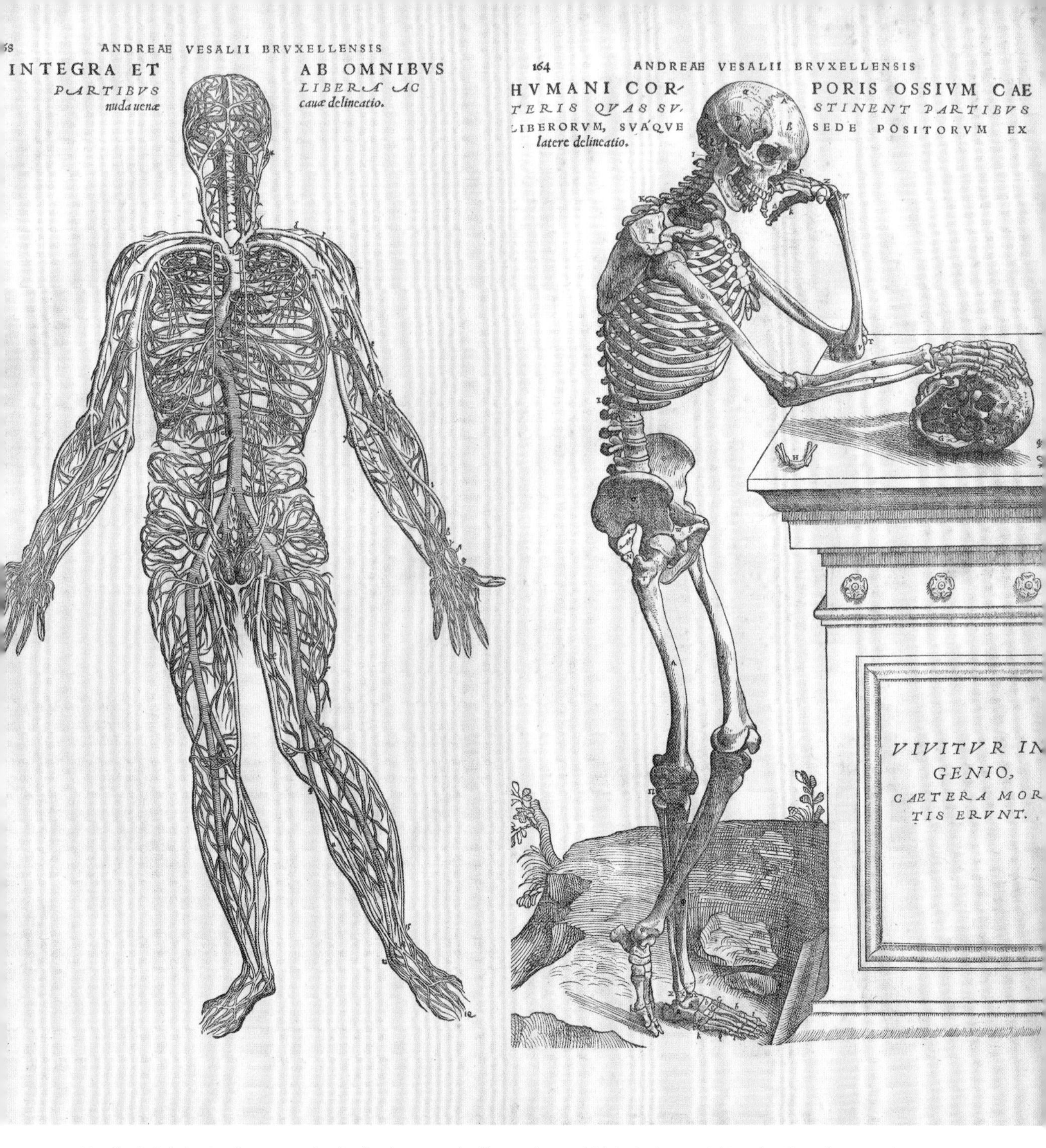

Vesalius's Fabrica *has become a classic, due in part to its illustrations, which include astonishing detail, as shown in the depiction of the circulatory system at left, and sometimes convey a wry wit, as shown in the pensive skeleton at right.*

Three Thinkers of the Renaissance

A Scientific Method Pioneer

A portrait of Caterina Sforza by Lorenzo di Credi, circa 1483

In a world before modern chemistry and biology, Caterina Sforza (1463–1509) is a Renaissance alchemist, which is someone who mixes chemicals in search of useful combinations. She collects hundreds of formulas for medicines and cosmetics, and compiles them in a book, *Experimenti*, which will be published after her death.

Sforza is a member of one of Italy's leading families. At a time when politics is turbulent and often deadly, she holds her own in that world, perhaps because she receives the same education as the boys in her family. And being an aristocrat gives her privileges.

Sforza is the kind of person sometimes described as "larger than life." She lives in a dynamic culture with more than its share of greed, thirst for power, and murder. Known as the Tigress of Forli, she raises her children on her own after her husband is assassinated, and she survives captivity in a dungeon. She rides horseback with elegance, handles a feud with the pope, establishes a scientific laboratory in her fortress palace (in the Italian town of Forli), writes a bestselling book, and creates beautiful gardens where she grows medicinal plants.

A salve she details in her book involves boiling meat and greens in wine for eight hours. It is intended to treat gout, a version of arthritis that causes severe joint pain. This would not be recommended by doctors today, but in her time, Sforza's book and her salves sell well. While most of her remedies seem to have little value today, and some may be dangerous, the big idea is that Sforza is experimenting, studying results, and then doing more experiments, which puts her among those pioneering the scientific method. Across Europe during this era of plagues, women are called on to be healers at a time when no one understands germs.

Is She a Scientist or a Doctor?

Being a midwife is a bit of both. Louise Bourgeois Boursier (1563–1636) is a Parisian midwife, which means she delivers babies. In her case, in Paris, that often means royal babies. One of the babies she delivers will become King Louis XIII.

Boursier has a scholarly mind and becomes an expert on the birthing process. By 1609, she has delivered two thousand babies. Most doctors ignore the placenta, an organ that holds and nourishes the growing baby in a mother's womb. Boursier, who writes a book about childbirth, describes the vital role of the placenta in the development of the unborn child. Her book becomes a big success and is translated into several European languages.

Louise Bourgeois Boursier, whose book about childbirth helped change obstetrics

Then a young mother whose baby Boursier delivered dies. Physicians who were unable to save the woman unfairly blame the midwife. Boursier will never deliver royal babies again. But she continues to work as a midwife, to write, to explain the birth process, and to support the skills of other midwives.

The Alchemist

In 1561, an Italian alchemist publishes a very successful book, *The Secrets of Isabella Cortese*. It is Cortese herself who has written and published the book, or so it is said. The book is filled with her recipes for medicines and cosmetics. One recipe for a face cleanser includes lemon, dried beans, white wine, eggs, and goat milk. Then as now, people want to look good and feel well, and Cortese's book seems to offer them formulas to do both.

But was there an Isabella Cortese? If so, there is little known about her. Meredith K. Ray, a modern author and Italian scholar, points out that if you mix up the letters in her name, you get the Italian word *secreto*, or "secret." Cortese even tells readers to burn her book after they learn its recipes.

VARIETÀ
DI SECRETI
DELLA SIGNORA
ISABELLA CORTESE,
Ne' quali si contengono cose Minerali,
Medecinali, Profumi, Belletti,
Artifitij, & Alchimia;
Con altre belle curiosità aggiunte.
DI NUOVO RISTAMPATI,
& con somma diligenza corretti.
IN VENETIA, MDCXIIII.
Appresso Lucio Spineda.

Isabella Cortese's 1561 book includes instructions for alchemy, turning base metals into gold, which can't be done, although many in her time were sure it could be.

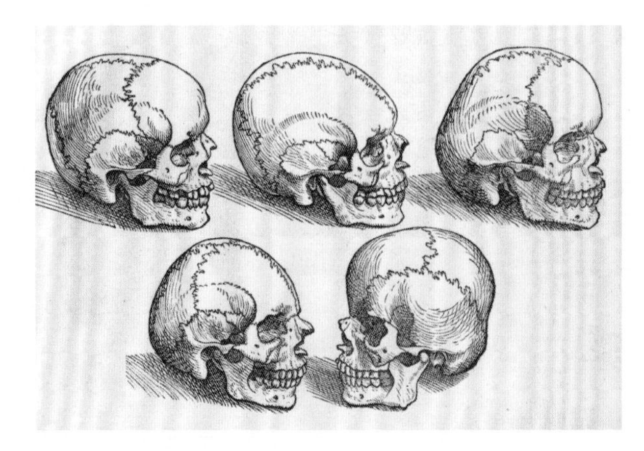

Human skulls from Vesalius's Fabrica

An exquisite textbook, the *Fabrica* is intended for a wider audience than medical students. It will be the first known book of anatomy based on dissections of real human cadavers. The illustrations in it are made using small woodblocks. Vesalius, a perfectionist, chooses Johannes Oporinus, the best printer in Basel, Switzerland, to print the book. Vesalius follows the *Fabrica* with the *Epitome,* meaning "summary," which is an abridged version of the *Fabrica* and includes eleven illustrations that focus on muscles. Later, many will call the publication of these books the beginning of modern life science.

The original blocks used for the *Fabrica* become part of a collection in the library at the University of Munich. During World War II, when the Allies bomb Munich, the blocks will be destroyed.

In the mid-1500s, the scholar-printer Johannes Oporinus publishes both the *Fabrica* and the world's first printed volume of the Qur'an in Latin.

In 1543, the same year that the *Fabrica* is printed, Copernicus, the Prussian mathematician and astronomer, publishes his great work on astronomy, *On the Revolutions of Celestial Spheres*. Copernicus has figured out, and explains in his book, that the planets, including Earth, take annual journeys around the sun. Before this, most experts thought the Earth was the unmoving center of the universe.

These two books—by Copernicus and Vesalius—become foundation stones for a scientific revolution that will begin a half century later when the Italian astronomer Galileo Galilei helps initiate a new kind of intellectual questioning: it starts with observation and continues with experimentation and proof. Modern science, which considers life by actually studying it, will emerge from this new thinking. Guenther will come to respect Vesalius, and eventually they will work together on a 1538 update of Guenther's textbook on life science, *Principles of Anatomy according to the Opinion of Galen*.

Now the story gets muddy. Vesalius moves to Spain, where someone—probably someone with clout—suggests that he make a trip to the Holy Land, or the region of the Middle East described in the Bible. A letter that still exists, written by a diplomat of the era, says that Vesalius got in trouble after mistakenly performing an autopsy on a patient who was not yet dead, and made a pilgrimage to the Holy Land as penance. But it is not known for sure if this is true. What is clear is that in 1564, his ship is destroyed during a storm in the Ionian Sea, and Vesalius dies on the island of Zakynthos. It is 1564 and he is forty-nine years old. ■

02 Cultures Clash, and a King Sends His Doctor to America

The Great Montezuma was about forty years old, of good height and well proportioned, slender, and spare of flesh. . . . He had good eyes and showed in his appearance and manner both tenderness and, when necessary, gravity. He was very neat and clean and bathed once every day. . . . For every meal, over thirty different dishes were prepared by his cooks. . . . Women brought him tortilla bread, and as soon as he began to eat they placed before him a sort of wooden screen painted over with gold. . . . From time to time they brought him, in cup-shaped vessels of pure gold, a certain drink made of cacao.

—Bernal Díaz del Castillo

How can we save our homes, my people?
The Aztecs are deserting the city:
The city is in flames, and all
Is darkness and destruction.

—from *Cantares mexicanos*

LEFT: These statues of warriors were created during the Toltec Empire, which lasted from the seventh century to the twelfth century. They stand in Tula, northwest of modern-day Mexico City.

Two American calendars: Aztec on the left, Nahuan on the right. Both reflect a sophisticated understanding of the year's 365-day cycle and include days that are named after deities. The circular calendar stone shown on the left measures about 12 feet (3.5 meters) across and weighs 27 tons (24.5 metric tons).

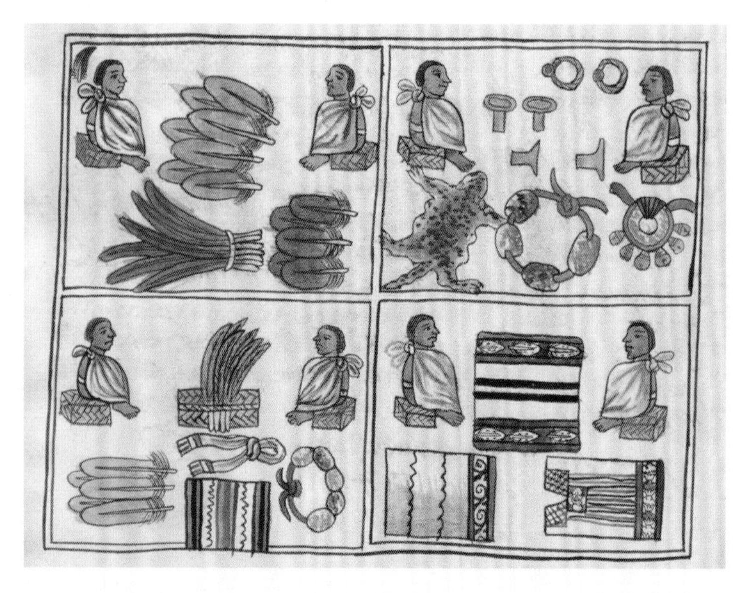

A page from the 1577 Florentine Codex*, compiled by a visiting Spanish friar, that shows Nahuan merchants selling feather articles, jewelry, textiles, and fur*

In 1500, the Aztecs are part of a Nahuan world of about six million people. They control a vast territory of hundreds of states, all paying tribute to the Aztec rulers. Some of those tributes are human beings, who are ritual sacrifices to their gods.

This great empire, which covers part of today's Mexico along with some adjacent areas, was founded in the first half of the fifteenth century; its greatest ruler, Montezuma, took power in 1440. Tenochtitlán (which Europeans will later name Mexico City) is the magnificent Aztec capital; when the Europeans arrive there, it has some 140,000 inhabitants. On market days, as many as fifty thousand people visit the city and support a prosperous economy. This civilization has developed writing and created books.

The sophisticated Aztec culture encourages the work of skilled artists, who create dazzling jewelry, statues, platters, and bowls of gold, silver, and gemstones. It is not prepared for the arrival of outsiders, though—especially outsiders with lethal weapons.

In 1519, when Hernán Cortés arrives in the Aztec Empire from Spain, he brings germs along with guns. Like most Europeans who have survived pandemics, his sailors have immunities to the viruses they carry; the Aztecs do not. Densely populated Tenochtitlán and

The oldest evidence of humans in the Americas goes back fifteen thousand to twenty thousand years, to a time often called prehistory (which means we don't have written documentation of what was going on then). The first people to arrive and settle in what is now South America may have come from Asia, across the Bering Land Bridge, but some anthropologists say they came from Polynesia, across the Pacific. There are several theories about settlements in the eastern part of the Americas.

A Nahuan censer, a device used during religious ceremonies to burn incense, in the form of the god Xochipilli

the surrounding towns are soon devastated. With much of the Indigenous population dying, Spanish soldiers can pillage with ease. They melt the precious treasures they find—the history and artistry of a culture—turning works of art into bars of gold and silver.

Some of the Indigenous people will be murdered by Spanish soldiers, who carry powerful weapons. The Spaniards enslave others, expecting them to do work no European will do. By 1521, the once awesome Aztec Empire is no more, though their presence

Smallpox victims depicted in the Florentine Codex. *An English translation reads that the disease brought "pustules that covered people" and "caused great desolation; very many people died of them, and many just starved to death. . . . Some lost an eye or were blinded."*

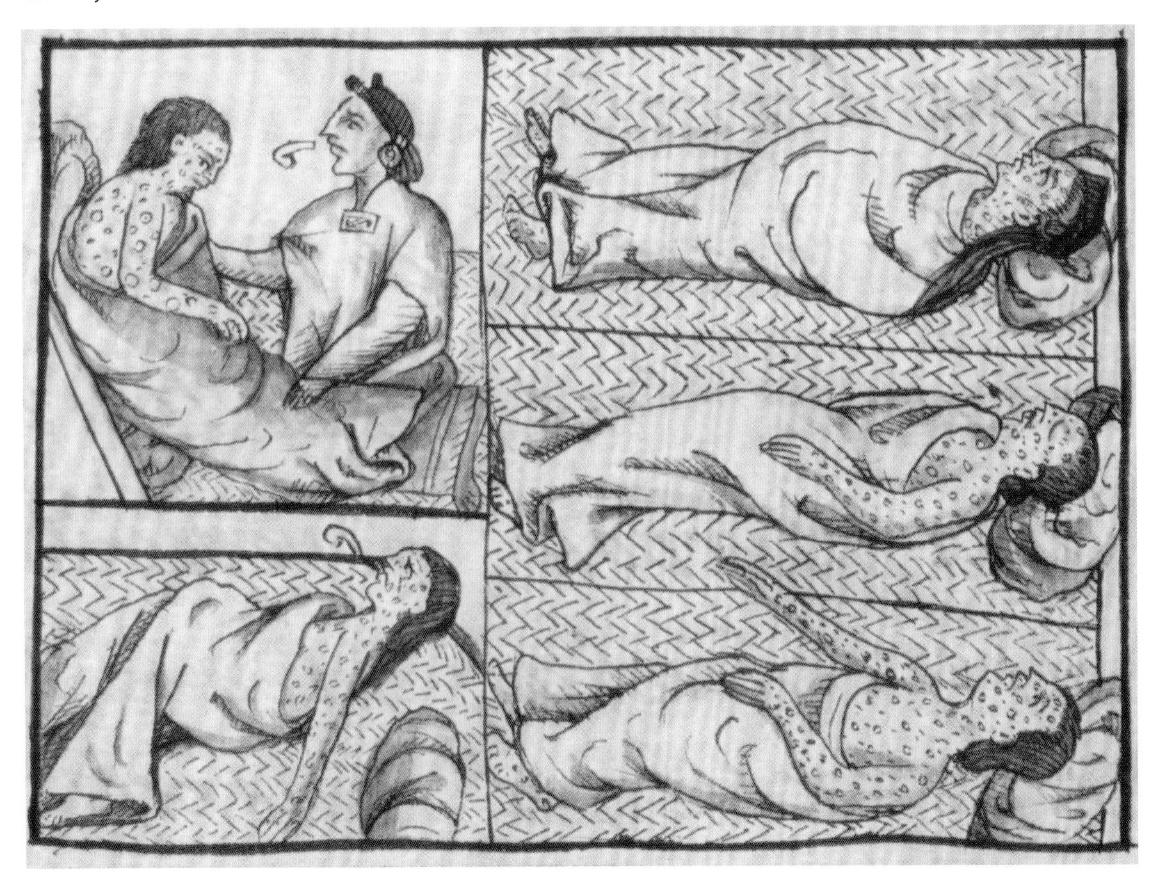

and influence remain, with Indigenous peoples making up about 15 percent of the Mexican population today.

In Spain, a few voices speak out about the atrocities in America. One of them belongs to Bartolomé de Las Casas (1484–1566), a priest who was among the early Spanish landowners in the Americas. Las Casas returns to Spain and writes *A Short Account of the Destruction of the Indies*, a chronicle of the first decades of European colonization in the West Indies. In it, he describes murder and barbarism committed by the Europeans. As an early European colonist, Las Casas encouraged enslaving Africans rather than Native Americans; then he came to realize the horrors of all forms of slavery. His is an informed voice when he returns to Spain, speaking out on Indies-related issues. In 1550, he takes part in what will be known as the famous Valladolid debate, in the Spanish city of Valladolid, where theologians and scholars debate how Indigenous peoples should be treated. Las Casas's opponents in that public debate degrade Native Americans, calling them uncivilized, less than human, and in need of Spanish rule.

An engraving from the Latin edition of A Short Account of the Destruction of the Indies *by Bartolomé de Las Casas. It shows Spanish troops meeting Indigenous peoples.*

Las Casas says the Native Americans are fully human and that enslaving them is wrong. "The reason why the Christians have killed and destroyed such an infinite number of souls is that they have been moved by their wish for gold and their desire to enrich themselves," he says.

In Peru, an Earth Shaker

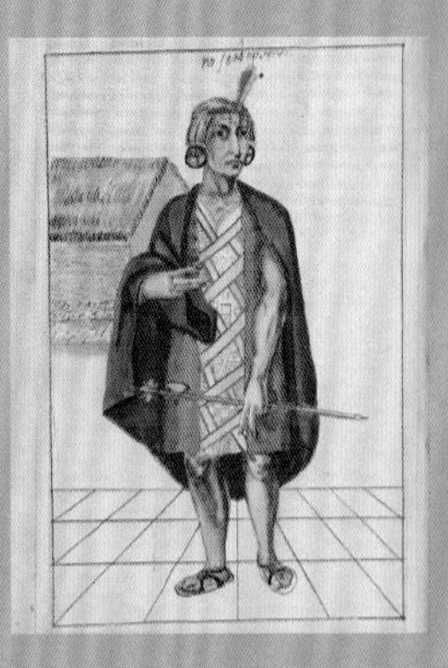

Pachacuti, depicted with a red feather in his hair

Pachacuti is an empire builder. He is one of the sons of a king in the 1400s, in the Andes Mountains among the Incan people in what is modern-day Peru. His full name is Pachacuti Inca Yupanqui, which means "he who shakes the earth with honor."

Pachacuti is not supposed to succeed his father. But after his hometown of Cusco is invaded by the neighboring Chanka people, his father flees. Pachacuti stays behind and organizes a successful defense of Cusco. He will soon supplant his father as the Sapa Inca, or king. The vast Incan Empire he builds will extend throughout much of the west coast of South America. Pachacuti is even believed to have built Machu Picchu, a famous mountain retreat where about a thousand people are thought to once have lived, an architectural marvel of stone buildings, precisely hewn windows, stunning views, and angular stairs.

Llamas can handle life on mountaintops. This image of a pack train at a high elevation in Peru was published in 1602.

Much of the empire, which he rules from 1438 to 1471, is in mountainous territory. The Incas carve terraced slopes into the land to plant corn, potatoes, and quinoa at different altitudes where they can flourish.

Incas use llamas to traverse their roads (they don't use wheeled vehicles), and they barter goods in intricate networks of exchange. They don't have a currency, so they rely heavily on bartering, and they don't have a written language, but they have a unique system of sending messages using long strands of corded knots called quipu.

In Incan belief, spirits can inhabit rocks, mountains, or streams. Even the dead are present; the Incas mummify the dead, and Pachacuti preserves the homes of his royal predecessors.

Machu Picchu, in Peru, elevation 7,970 feet (2,430 meters), or 1½ miles (2.5 kilometers), above sea level

Building an empire can be a bloody business. Pachacuti forces thousands of vanquished foes to relocate to the far reaches of his dominion. The empire that Pachacuti builds falls in the years after his death due to the arrival of Spanish conquerors and the smallpox they bring along with them, as well as civil war. But its legacy is central to the region's cultural history.

A llama in front of the Machu Picchu archaeological site

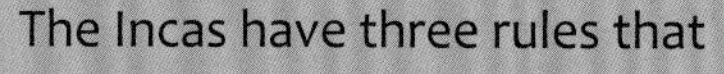

The Incas have three rules that stand out above others: do not steal, do not lie, do not be lazy. If you follow the rules, you get to go to the land of the sun after you die and bask in warmth. If you break the rules, you go to freeze in the underworld. And if you're somewhere in between, like most people, you are reborn on Earth and try again. Do you follow the Incan rules?

In Spain, the devout King Philip II (1527–1598) sees the Bible as much more than a guide to issues of heart and soul. He sees it as a book of history and geography, and a literal description of the world as it was when it was created, as many other Europeans do. Philip believes it is God who has given him his victories and that it is his job to bring Christ to the native people of America. He sends priests to convert them.

In 1571, Philip II also sends his personal physician, Francisco Hernández de Toledo (1514–1587), to the Americas to search for new medicines and to explore the flora and fauna of the region. Philip II has been getting very rich on American gold, and he wants to know more about the lands he has conquered. Sending a scholarly scientist makes sense. King Philip has chosen well. Hernández is among the Renaissance doctors practicing a new kind of medicine based on classical ideas from the ancients, and he also has translated a work on life science by Pliny the Elder, an ancient Roman philosopher and naturalist. Hernández will work hard at his new assignment.

Imagine life before photography. If you can afford it, you may go to an artist and have your picture painted. If you are royalty, you can pick among the best artists of your time. Philip II chose Titian (1488–1576), one of the greatest artists in history, to paint his official royal portrait, but what you see here is a replica, which is housed in Madrid's Museo Nacional del Prado. Note that Philip II is wearing the chain of the Order of the Golden Fleece.

King Philip II of Spain

Hernández sets sail in August 1571. Given Philip's royal commission, Hernández crosses the ocean, taking a geographer, a botanist, and his son Juan with him. This is the first scientific expedition from Europe to America.

Illustrations from the Florentine Codex *depicting Nahuan warriors (left) and rituals (right)*

Hernández will spend seven years in Central America and the Valley of Mexico, studying native plants and Indigenous medical practices. He will work with herbalists and scholars of the Nahuan culture and document everything he can find. He will even eat tortillas.

When Hernández returns to Europe, it is with twenty notebooks filled with his observations as well as detailed drawings done by three Indigenous artists he has baptized. Among the drawings are renderings of native plants unknown to Europeans, such as vanilla, corn, cacao, tobacco, chilis, tomatoes, and cacti. His written accounts of the

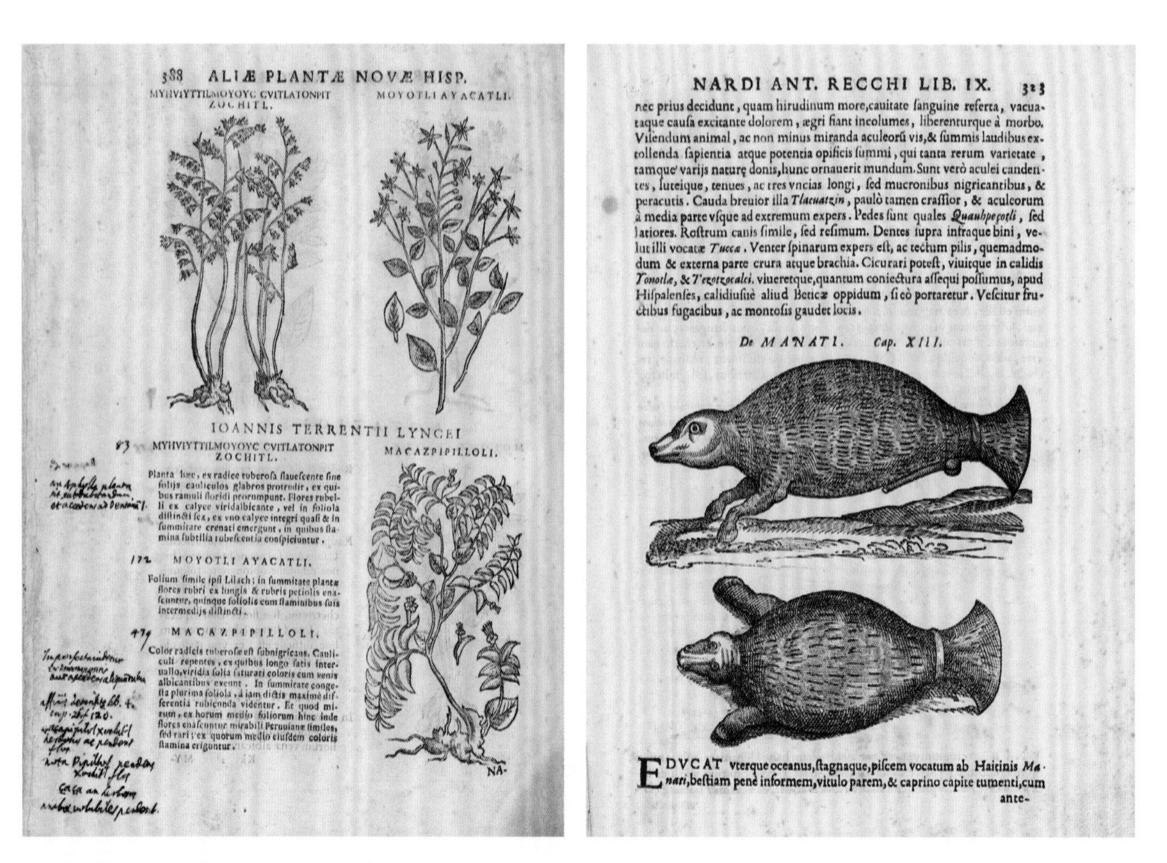
388 ALIÆ PLANTÆ NOVÆ HISP.

MYHVIYTTILMOYOYC CVITLATONPIT ZOCHITL.

MOYOTLI AYACATLI.

IOANNIS TERRENTII LYNCEI

MYHVIYTTILMOYOYC CVITLATONPIT ZOCHITL.

Planta hæc, ex radice tuberoſa flaueſcente ſine folijs cauliculos glabros protrudit, ex quibus ramuli floridi prorumpunt. Flores rubelli ex calyce viridalbicante, vel in foliola diſtincti ſex, ex vno calyce integri quaſi & in ſummitate crenati emergunt, in quibus ſtamina ſubtilia rubeſcentia conſpiciuntur.

MOYOTLI AYACATLI.

Folium ſimile ipſi Lilach; in ſummitate plantæ flores rubri ex longis & rubris petiolis enaſcuntur, quinque foliolis cum ſtaminibus ſuis intermedijs diſtincti.

MACAZPIPILLOLI.

Color radicis tuberoſæ eſt ſubnigricans. Cauliculi repentes, ex quibus longo ſatis interuallo, viridia folia ſaturati coloris cum venis albicantibus exeunt. In ſummitate congeſta plurima foliola, à iam dictis maximè differentia rubiçunda videntur. Et quod mirum, ex horum medio foliorum hinc inde flores enaſcuntur mirabili Peruuianæ ſimiles, ſed rari; ex quorum medio eiuſdem coloris ſtamina eriguntur.

NARDI ANT. RECCHI LIB. IX. 323

nec prius decidunt, quam hirudinum more, cauitate ſanguine referta, vacuataque cauſa excitante dolorem, ægri fiant incolumes, liberenturque à morbo. Viſendum animal, ac non minus miranda aculeorũ vis, & ſummis laudibus extollenda ſapientia atque potentia opificis ſummi, qui tanta rerum varietate, tamque varijs naturę donis, hunc ornauerit mundum. Sunt verò aculei candentes, luteique, tenues, ac tres vncias longi, ſed mucronibus nigricantibus, & peracutis. Cauda breuior illa *Tlacuatzin*, paulò tamen craſſior, & aculeorum à media parte vſque ad extremum expers. Pedes ſunt quales *Quauhpeçotli*, ſed latiores. Roſtrum canis ſimile, ſed reſimum. Dentes ſupra infraque bini, velut illi vocatæ *Tucca*. Venter ſpinarum expers eſt, ac tectum pilis, quemadmodum & externa parte crura atque brachia. Cicurari poteſt, viuitque in calidis *Tonotlæ*, & *Tezotzocalci*. viueretque, quantum coniectura aſſequi poſſumus, apud Hiſpalenſes, calidiuſuè aliud Beticæ oppidum, ſi eò portaretur. Veſcitur fructibus fugacibus, ac montoſis gaudet locis.

De MANATI. Cap. XIII.

EDVCAT vterque oceanus, ſtagnaque, piſcem vocatum ab Haitinis *Manati*, beſtiam penè informem, vitulo parem, & caprino capite tumenti, cum ante-

Pages from an annotated compilation of Francisco Hernández's observations and commissioned drawings published in 1651 showing plants (left) and a manatee (right)

plants are often long and detailed; he describes four varieties of cacao alone. He also observed insects, birds, mammals, and reptiles—including the Gila monster.

Hernández brings home pineapple, cocoa, and corn—all unfamiliar in Europe—along with boxes and boxes of seeds and plants. His intent is to compile a massive reference compendium to identify and describe all the plants and animals of South America. His natural history of New Spain encompasses six volumes of text and another ten with illustrations. Written in Latin, it details more than three thousand plants. But Hernández falls out of favor in the royal court and is replaced as royal physician; his work is not published in his lifetime.

Hernández has acquired priestly training as well as medical knowledge in the course of his studies in Spain. That explains why he is able to baptize some of the Indigenous people who work with him. He believes he is saving their souls when he does so.

He dies in 1587, and his notes and materials are placed in El Escorial, Spain's royal residence, where they are consulted by a generation of scientists, medical specialists, and natural philosophers. And then, in 1671, much of his work is destroyed by a fire, though remnants of it are published.

As for King Philip II, he no longer has time to consider nature in a faraway land. A political drama is unfolding in Europe.

Sometimes called Philip the Prudent, Philip II is not just Spain's ruler; he is king of Naples and Sicily, Duke of Milan, an archduke of the German House of Habsburg, and ruler of the Netherlands and the Philippine Islands, which are named for him. The phrase "the empire on which the sun never sets" was coined in the day of Philip's father, Charles V,

El Escorial, the royal residence near Madrid (top); the library at El Escorial (bottom)

to describe their empire, which includes land on every continent known to Europeans (they don't yet know about Australia or Antarctica). Given his access to weapons and soldiers, Philip believes he can take whatever he wants from the 1.3 million Indigenous people who live in South America's Andes Mountain region.

A small man with a protruding chin, Philip hopes to marry Queen Elizabeth I; it would secure his right to rule over England. He has religious motives too. Philip is the force behind the Spanish branch of the Inquisition, which he deeply believes is doing God's work when it murders those who refuse to follow the Catholic faith.

The Inquisition was a tribunal within the Catholic Church set up by Pope Gregory IX in 1232 to root out and punish heresy, as church officials defined it. It continued in Europe and then in the Americas for hundreds of years and was known for the severity of its tortures and its terrorizing and persecution of Protestants, Jews, Muslims, and anyone who had ideas that challenged mainstream Catholicism or Spanish rule.

Queen Elizabeth I (1533–1603) is not Catholic. Her father, Henry VIII, took England out of the Catholic world and created the Protestant Church of England. As queen (from 1558 to 1603), Elizabeth is leader of that church, which is also called the Anglican Church. Philip, a devout Catholic, intends to bring England back into the Catholic fold by marrying her.

Elizabeth, who is shrewd, doesn't say yes and doesn't say no.

I will be as good unto ye as ever a Queen was unto her people. No will in me can lack, neither do I trust shall there lack any power. And persuade yourselves that for the safety and quietness of you all I will not spare if need be to spend my blood.

—Queen Elizabeth I

At the same time, pirate ships are attacking gold-laden Spanish ships crossing the Atlantic, and England's queen decides to support them by granting them official royal recognition as privateers, a term for privately owned armed ships that are commissioned by a government. This designation means they won't be treated as pirates, who operate outside any nation's laws.

These privateers become known as sea dogs. Her move riles Philip II, a powerful monarch who is used to having whatever he wants. He becomes obsessed with England and its queen.

By the late 1580s, Philip has given up the idea of marriage and makes plans to send roughly 130 to 150 ships north to England. With that large group of ships—an armada—he intends to wipe out Britain's inferior fleet and set the stage for a Spanish invasion of England. This is a bold plan from the world's most formidable ruler.

Queen Elizabeth is not about to wait for Philip to get his invasion force in order. She encourages the daring pirate Francis Drake to lead a small fleet to the Spanish port of Cádiz, where some of the Spanish ships have gathered.

Drake destroys the ships and tons of supplies. He laughingly calls his attack the "singeing of the king of Spain's beard." It delays the launch of Spain's great fleet for a year, and that gives the English time to get ready.

Elizabeth's soldiers prepare for an invasion of England: they build trenches and hang a huge metal chain across an estuary of the Thames River. The queen sees that forty warships are readied and that long-range guns are put on merchant vessels. Then Queen Elizabeth herself, dressed in a white velvet gown with military

Francis Drake

The Spanish Armada as depicted in a 1612 book

The Armada Portrait of Elizabeth I, Queen of England and Ireland, circa 1588. Scenes of the battle are behind her.

This life-size painting of Elizabeth I was created about 1588, and there is some question as to who painted it. It is unusual in portraiture of the time in England because it includes other pictures embedded in the portrait. Elizabeth has turned her back on storm and darkness and is gazing where there is sunshine. This clearly is a celebration of the defeat of the Spanish Armada. There are three versions of this portrait, and experts say they may be the work of three different artists: one is at Woburn Abbey, a cut-down version is at the National Portrait Gallery in London, and one is privately owned. Three paintings of the queen and the armada? Yes—this was a huge moment in British history, and these paintings are a form of celebration.

In the background scenes, you can see triumphant and brightly lit English ships and Spanish ships in distress and gloom.

Some say the mermaid carved in the chair (to the right) represents the executed Queen Mary. Note Queen Elizabeth's mouth. She isn't about to smile and show you her teeth. Elizabeth, being rich, was able to eat a lot of sweets. Her teeth became a mess of dark, painful rot. Her dentist, suggesting the latest dental remedy of the times, told her to rub sugar on her teeth.

trimmings, speaks to her troops as they prepare for battle. Her words will be repeated across the land; here is some of what Elizabeth says:

> *I know I have the body of a weak, feeble woman; but I have the heart and stomach of a king, and of a king of England too, and think foul scorn that Parma or Spain, or any prince of Europe, should dare to invade the borders of my realm; to which rather than any dishonor shall grow by me, I myself will take up arms, I myself will be your general, judge, and rewarder of every one of your virtues in the field.*

Philip's fleet is the largest ever known in Europe. Spain's king believes that God is on his side, so he doesn't consider the queen's determination, or weather issues, or the agility and power of the small English ships, or luck, or disease: all of which will contribute to undoing his armada. Spain loses roughly half of its ships and some fifteen thousand sailors and soldiers in the attempted invasion, though estimates vary.

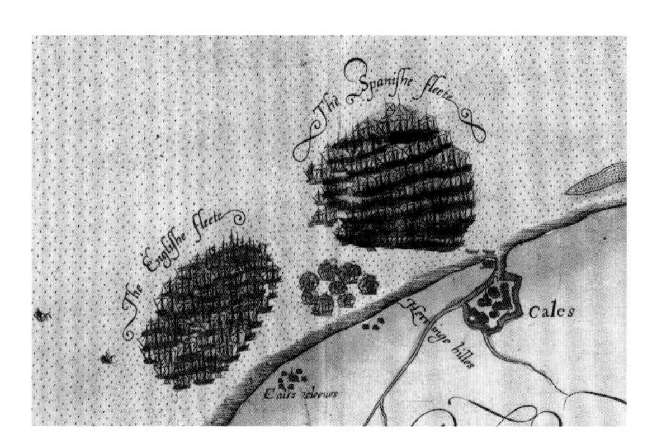

A map detail of the battle between England and the Spanish Armada

The destruction of Spain's armada in 1588 is a huge and unexpected victory for England and its monarch. Elizabeth I, a female ruler, won a big battle against the world's most powerful monarch and changed the trajectory of history. When William Shakespeare writes the play *A Midsummer Night's Dream* (in 1595 or thereabouts), he says, "And though she be but little, she is fierce," and most people are quite sure he is thinking about England's queen.

While this history-changing fight is commanding attention in Europe, hardly anyone is thinking about the scientific exploration of the Americas conducted by Hernández. His notes are mostly forgotten, except by a scholarly young man in Italy. ■

03

The Sharp-Eyed Lynxes Want to Know More: One of Them Is Named Galileo

I do not feel obliged to believe that the same God who has endowed us with senses, reason, and intellect has intended to forgo their use and by some other means to give us knowledge which we can attain by them.

—Galileo Galilei

Scientists who become icons must not only be geniuses, but also performers, playing to the crowd and enjoying public acclaim.

—Freeman Dyson

LEFT: This statue of Galileo is outside the Carnegie Museum of Natural History in Pittsburgh, Pennsylvania. It was commissioned by Andrew Carnegie, an industrialist who did not want statues of dinosaurs outside his museum; he wanted works that celebrate human creativity and achievement in art, science, literature, and performance.

Philip II dies in 1598 and Queen Elizabeth I in 1603, ending a political era in Europe. The Western world is soon teetering between a known past, dominated by a powerful church and a few superstar families, and new questions being asked at the growing universities. Copernicus's big idea—that the Earth revolves around the sun and not the other way around— suggests a new order to the universe. Meanwhile, the anatomical drawings in Vesalius's marvelous book the *Fabrica* have some professors and their students carefully examining life-forms. Perhaps life and its workings can be understood by these "modern" people who live in the Renaissance and beyond. Perhaps scholars don't have to depend on writings from the ancient world.

William Shakespeare, the great playwright, sometimes known as "the Bard"

"The Heavens," the stage ceiling at the modern Globe Theatre in London, a reconstruction of the original, where Shakespeare's plays were performed in his time

Because of a proliferation of printing presses, some astonishing writers are soon bringing big ideas to the reading audience, which keeps growing. Shakespeare, a playwright and poet, is the writer many want to read. But Aristotle, Miguel de Cervantes, Dante Alighieri, Desiderius Erasmus, Michel de Montaigne, Niccolò Machiavelli, and Geoffrey Chaucer are also all writing compelling works (then and now) in English, Spanish, Italian, and other languages.

Does Shakespeare, who lived from 1564 to 1616, know about the scientific revolution unfolding in his time? Yes. Shakespeare is well aware of the groundbreaking work of his contemporary Galileo Galilei (1564–1642) and also of the British astronomers who are studying the night sky that now, given the magnification of some telescopes, is showing them previously unknown stars. At a time when many of his peers turn to astrology for judgments and predictions, Shakespeare doesn't consult the stars. In Sonnet 14, he writes:

Two engravings of Galileo; in the one on the right, he offers a telescope to three women, possibly Urania, the Greek goddess of astronomy, and her attendants, while he points toward the sky, where some of his discoveries are depicted.

> *Not from the stars do I my judgment pluck,*
> *And yet methinks I have astronomy;*
> *But not to tell of good or evil luck,*
> *Of plagues, of deaths, or seasons' quality.*

While he doesn't mention Copernicus or Galileo, Shakespeare's writing is suffused with astronomical turns of phrase. One example comes in his drama *Othello*. The title character, speaking of the moon's uneven orbits, says, "It is the very error of the moon, / She comes more nearer earth than she was wont / And makes men mad."

Shakespeare is at the peak of his creative power in 1595, when an influential atlas by Gerardus Mercator (1512–1594), a Flemish cartographer, is published after Mercator's death. Another contemporary, Thomas Harriot, a mathematician-astronomer, becomes known for using telescopes to study England's night sky. One of Shakespeare's neighbors, Thomas Digges (1545–1595), writes about the "new astronomy." We can be sure that Shakespeare knows of these men.

In one of his most famous plays, *Hamlet*, Shakespeare chooses an actual Danish castle as the setting, which he refers to as Elsinore. The real castle sits within sight

A portrait of Flemish cartographer Gerardus Mercator (left) and pages from his 1595 atlas (right)

of the famous astronomer Tycho Brahe's observatory. Two of the play's characters, Rosencrantz and Guildenstern, are named for relatives of Brahe.

The medieval world is giving way to the Renaissance, and those keeping up with the times are aware that science is where intellectual excitement can be found. When Shakespeare's character Julius Caesar says, "I am as constant as the northern star," the playwright is showing that in his world (which has no satellites or radar), everyone is aware of the navigational usefulness of that star. Might he be thinking of Copernicus and Vesalius and other scientific innovators when he gives Hamlet these

Tycho Brahe, a Danish noble, was an astronomer and physicist who built an observatory in his castle that was said to be the best in Europe.

lines: "There are more things in heaven and earth, Horatio, / Than are dreamt of in your philosophy"? The message for his audience seems to be: if you are not keeping up with the new scientific ideas, you are missing out. And in the play *Antony and Cleopatra*, Shakespeare admits to his own interest in the natural world: "In nature's infinite book of secrecy / A little I can read."

The new ideas, and the questions they raise, are in some cases supported by people of wealth, such as the Medicis, the most powerful family in Italy. (Throughout history, wealth and influence play a huge role in who has the freedom to ask creative questions.)

The energetic and innovative era called the Renaissance began in Italy in the fourteenth century but is still in full swing in 1603, when Federico Cesi (1585–1630), a well-off teenager who lives near Rome, becomes a question asker.

Federico Cesi

Cesi isn't interested in Spain's armada or in political power. A true scholar as well as the lord of a town north of Rome called Acquasparta, he wants to know what Hernández discovered in the Americas, and he has the means to find out. Cesi is part of an affluent, educated elite that includes women, like his intelligent mother, who have the time, the taste, and the means to pursue great art, good writing, and ideas from the scholarly world. The intellectual and writer Margherita Sarrocchi is among his friends.

Almost everyone worries about offending church leaders, but Cesi is an insider. His uncle is a cardinal in the Roman Catholic Church; that gives him rare privileges and the security to follow his heart and his mind. Aware of new thinking that is expanding outlooks during the Renaissance, he is eager to put aside the ancients and study nature itself.

Margherita Sarrocchi (1560–1617) inhabits the same social world as Cesi and Galileo. She is born into a well-to-do family in Naples, and it is clear at an early age that she is unusually bright—a child prodigy, in fact. Her family sees to it that she has a classical education that includes the sciences. Sarrocchi is taught by some of the most respected scholars of her time at the convent of Santa Cecilia in Rome; there she learns Greek and Latin, logic, geometry, astrology, mathematics, and theology. When she is fifteen, Rome's leading literary publication includes a poem she has written.

Sarrocchi marries and moves to Rome. Both a mathematician and a poet, she hosts gatherings, called salons, in her new home, where she brings together fellow poets and mathematicians, as well as other well-regarded thinkers. Her home is described as "the meeting place and academy for the best minds in Rome." Galileo is part of her circle of intellectual friends, along with Cesi and the mathematician Luca Valerio. When Galileo leaves Rome, he and Margherita Sarrocchi write to each other.

Sarrocchi writes an epic poem. It tells the story of a European warrior prince who fights against an Ottoman sultan. Filled with battles and romance, it is unlike any work by a European woman that has been published before.

La Scanderbeide, the book Sarrocchi is remembered for today, is published after she dies.

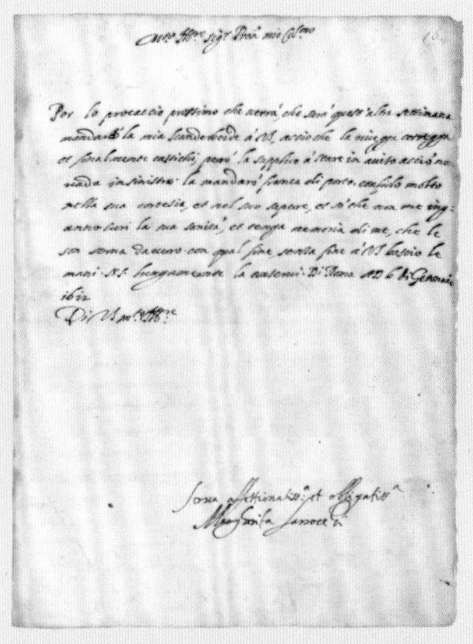

LA SCANDERBEIDE
POEMA HEROICO
DELLA
SIG.RA MARGHERITA
SARROCCHI
ALLA PRINCIPESSA D. GIVLIA DA ESTE.
DAL SIG. GIOVANNI LATINI
Con Priuilegio della Santità di Nostro Signore, del Rè di Spagna, e de' Prencipi d'Italia, dato alla Stampa.
IN ROMA, Per Andrea Fei. MDCXXIII.
CON LICENZA DE' SVPERIORI.

A letter from Sarrocchi to Galileo (top); the title page of her book La Scanderbeide, *a long poem about an ancient hero (bottom)*

Polymaths are people who know a whole lot about a whole lot. These are not people with superficial knowledge; they have minds that retain what they learn. The knowledge of a polymath spans different fields of study. Leonardo da Vinci is a classic example of a polymath; he was not just an artist but also an accomplished medical scholar, inventor, mathematician, and engineer.

This lynx stamp is from a book in the collection of Federico Cesi.

RIGHT: An engraving by Matthias Greuter, a member of the Lynxes, shows anatomical details of bees as seen through a microscope.

Cesi is eighteen in 1603, when he invites three scholars to join him to study Hernández's notes from the Americas. The trio are Johannes van Heeck, a Dutch physician; Francesco Stelluti, a mathematician; and Anastasio de Filiis, a scholar and polymath. Both Stelluti and de Filiis are from Umbria, which is in the middle of the Italian boot.

Cesi expects that he and his colleagues will study Hernández's notes and then publish and share what Hernández learned about the land across the great ocean. Cesi is interested in details about life there, along with information on plants, animals, and medications that may be unknown to Europeans. Calling themselves the Accademia dei Lincei (Academy of the Lynxes), these four intellects establish what most of today's historians agree is the first scientific society in the Western world. In a founding credo, they task themselves to "not only acquire knowledge of things and wisdom" but also "display them to men, orally and in writing, without any harm." In that credo, they announce that they will "be slaves neither of Aristotle nor any other philosopher." In other words, they intend to examine and consider the world about them without regard to what others may have said before them.

VRBANO VIII·PONT·OPT·MAX
Cùm accuratior ΜΕΛΙΣΣΟΓΡΑΦΙΑ,
à LYNCEORVM ACADEMIA,
in PERPETVAE DEVOTIONIS Symbolum
ipsi offerretur.
Magne Parens rerum, cui se Natura volentem
Subijcit, et dominos collocat ante pedes,
Respice, Naturâ quâ nil præstantiº omni,
E BARBERINAE stemate Gentis APEM:
Hãc vti Lyncëidum, ꝑpiori lumine lustrans,
Disposuit Tabulis, explicuitq; labor.
Cæsiadę Genio sacrum stimulante laborem,
Palladis et promptos arte iuuante viros,
Maxima dum tereti surgunt miracula vit
Maioremq; oculus discit habere fide
Quis nôrat quinas Hybl̨eo in corpore lingu
Atque leoninæ proxima colla iubæ,
Hirsutósq; oculos, binasque labris vagina
Nî facerent artis dīa reperta noua
Sic decet, vt dum te mirādū suspicit Orbi
Et magè mirandā se Tua præstet APIS
Ponebat IVSTVS·RIQVIVS·LYNCEVS·BELGA DEDIC·S·E·
ROMÆ
Superior. permissu 1625
FRANCISCVS STELLVTVS LYNCEVS FABR: IS MICROSCOPIO obseruabat
M. Greuter
delineau. incid.

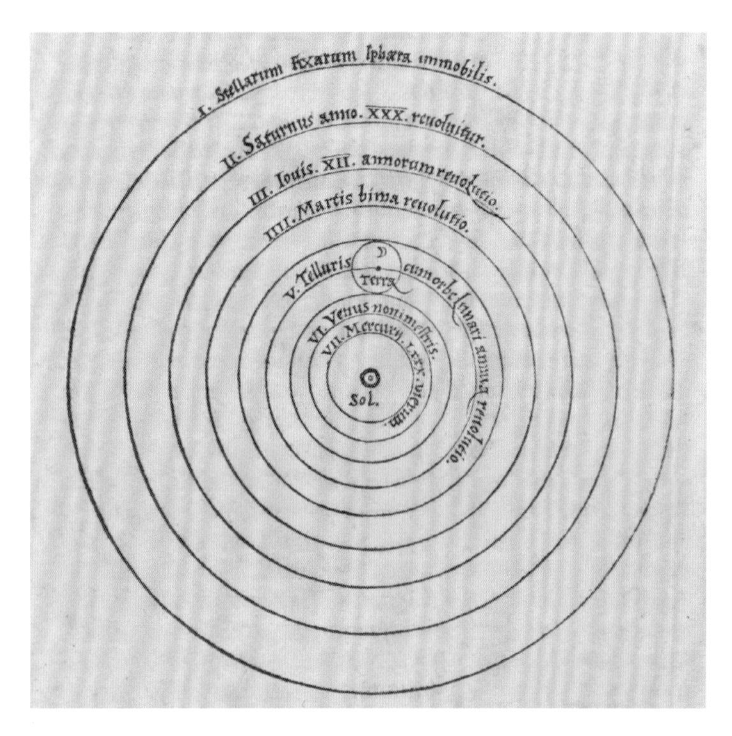

A diagram of the solar system by Nicolaus Copernicus, with the sun in the center instead of the Earth. Although that placement seems obvious today, at the time it created a whole lot of fuss.

The quartet chooses the lynx as their symbol because that wildcat has sharp eyes. They make plans to identify and classify all of nature, including all flora and fauna—a huge task, to say the least.

The Lynxes have been influenced by two controversial masterworks published in 1543: Copernicus's great astronomical work, *On the Revolution of Celestial Spheres*, and Vesalius's illustrated book on the human body, the *Fabrica*. Both books are on the Catholic Church's forbidden reading list. This is no little thing: the Inquisition has been burning heretics alive for reading books from that list. And yet those forbidden books are passed from one reader to another. Most thinkers want to know what they say.

In 1611, the Lynxes have been thriving for eight years when Cesi hosts a magnificent banquet to honor a newly selected member: Galileo Galilei, who is dazzling those of his time with his telescope and its startling images of the heavens. Galileo, with one of the world's greatest scientific minds, becomes the elite group's intellectual leader. Clearly proud of the association, he uses *Galileo Galilei Linceo* as a signature.

For Galileo, it isn't enough to think about what will happen when you drop two balls of different weights; you have to actually do it, more than once. It will be a while before the idea of controlled experiments takes hold, but this is a beginning. In Cesi, Galileo has found a sympathetic compatriot.

In 1624, Galileo sends Cesi a gift. This note goes with it:

Most Illustrious and Excellent Lord . . .

I am sending Your Excellency an "occhialino" to see the tiniest things close up. . . . I delayed in sending it because I had not yet brought it to perfection, having had difficulty in finding a way of grinding the lenses perfectly. The object is attached to the mobile circle, which is in the base, and is moved about so as to see all of it.

The device he calls an *occhialino* has three biconvex lenses. It magnifies about thirty times, the same as Galileo's telescopes. Today we know it as a microscope, but that term won't be coined until 1625 (by Giovanni Faber, another member of the Lynxes). With this device, Galileo introduces his patron to an unknown world, which will turn out to be much larger than anything Hernández found in Mexico.

Galileo describes what he has seen using his *occhialino*:

I have contemplated a great many tiny animals with infinite admiration. Among them the flea is most horrible, the mosquito and the moth are beautiful; and with great satisfaction I have seen how it is that flies and other tiny creatures can walk attached to mirrors, and even upside down. . . . In brief, one can contemplate infinitely the grandeur of nature, and how subtly she works, and with what indescribable diligence.

Galileo is more interested in the multitude of stars he can see through his telescope than in the insects he examines

None of the actual microscopes that Galileo used have survived. The microscope pictured here was perhaps designed by Galileo; it may have been built by Giuseppe Campani (1635–1715), a well-known watchmaker and lens grinder. It is now on display in Bethesda, Maryland, at the National Institutes of Health.

A replica of Galileo's microscope

The Great Comet of 1577, which was visible across Europe and which the astronomer Johannes Kepler witnessed as a child

Johannes Kepler (1571–1630), a German astronomer who lives in the same era as Galileo, has ideas that will help reinvent physics and our understanding of the movement of planets. Galileo writes to him, saying:

> *My dear Kepler, what would you say of the learned here, who, replete with the pertinacity of the asp, have steadfastly refused to cast a glance through the telescope? What shall we make of this? Shall we laugh, or shall we cry?*

close-up with his microscope. But several of the Lynxes are eager to try Galileo's magnifying device, and they will use it to make and publish detailed drawings of plants and fungi and insects. They also publish an edited and beautifully illustrated edition of part of Hernández's work; it is in Latin, which means it can be shared in Europe's scholarly community. Before long, there are thirty-two Lynxes. They are pioneering a new way of thinking, and they are aware and proud of that.

Some in this time believe that all of science's important questions have been answered, by either Aristotle or the Bible. So why should anyone bother with this field? Cesi's aristocratic father doesn't want his son dabbling in what seems like unproductive work. But Cesi's mother, Olimpia Orsini, who hails from an influential family, thinks differently and backs her son.

Still, dark rumors spread: the Lynxes are accused of black magic, in an era when witch trials take place in Europe.

At the same time, some church leaders are less than enthusiastic about new ideas. And the Inquisition gets increasingly powerful and despotic. Everyone, especially Galileo, knows what the Inquisition did to the great thinker Giordano Bruno, who, in 1600, was burned naked and upside down on a pyre in Rome for expressing ideas that today might be called "scientific."

Cesi offers his intellectual friends, like Galileo, some protection from the Inquisition, but not forever.

Unexpectedly, after a brief high fever, Cesi dies. He is forty-five years old.

The Lynxes are not only devastated; they are also without a patron and protector. Galileo expected the Lynxes to pay for publication of his masterwork, *Dialogue Concerning the Two Chief World Systems*. Now he is without a publisher. The group disbands.

The years ahead will be difficult for Galileo. Back in 1610, he wrote a book called *The Starry Messenger*; it confirmed that Copernicus was right. Earth does go around the sun. And his book *Dialogue Concerning the Two Chief World Systems*, which is published in 1632, further explains the action in the heavens. He writes this book in Italian, not scholarly Latin, and it is widely read.

The pope and the Roman Catholic Church are not pleased. If our Earth is not the center of the universe, doesn't that contradict the Earth-centered perspective of the Bible?

Galileo is in trouble. His books are banned. He faces a trial for his ideas, and he is convicted of heresy and sentenced to prison, which is changed to house arrest. For the last nine years of his life, one of history's greatest scientists is unable to leave his home.

More than three hundred years after his trial, in 1992, the Catholic Church formally clears Galileo of wrongdoing. It now agrees: the sun, not Earth, is the center of our solar system. But that doesn't help Galileo in his own lifetime. As for the Lynxes and the idea of independent scientific research, an organization carrying its name will be

An 1847 painting of the trial of Galileo. His trial and his fight with church authorities eventually helped lead to more intellectual freedom in the Western world.

Women Speak Up

It is 1599, and Giuseppe Passi, who lives in Venice, publishes a book called *The Defects of Women.* Passi says women are vain and cruel and can't be relied on. He calls them a "necessary evil."

Quite a few women in Venice can read, so it doesn't take long for Passi to get a powerful response. The next year, Lucrezia Marinella (1571–1653), who is a well-regarded (and brilliant) poet and novelist, publishes a book titled *The Nobility and Excellence of Women and the Defects and Vices of Men.* Marinella's tongue-in-cheek book is a step-by-step rebuttal of Passi; it sells very well and soon has second and third printings.

Lucrezia Marinella in a 1652 portrait by Giacomo Piccini

Marinella, speaking for many women, says, "No firmer stability than theirs can be found, as is revealed in the great patience they show in bringing up, feeding, and teaching the impatient male, which is something to marvel at."

Then Marinella gets serious and provides examples through time of learned and accomplished women. She writes:

> *Some people, possessing little knowledge of history, believe that women who are skilled and learned in the arts and sciences have never existed and do not exist now. To them such a thing appears impossible, nor can they yet understand that they see and listen to these women every day. These people have persuaded themselves that Jove has bestowed wit and intellect only on men and left women deprived of even the smallest quantity.*

The reality, as Marinella makes clear, is that women in the past have almost never been given opportunities equal to those that men routinely receive. Typically, women get little education and are not allowed to attend universities. Those who do get an education usually are daughters of the wealthy and have private tutors.

Marinella writes that if women were able to delve fully into the sciences and the "military arts," then "what would men do all day but raise their eyebrows and remain stupefied and full of admiration?"

Marinella is the well-educated daughter and wife of physicians. In her book, she rebuts critics of women going back to Aristotle. "Perhaps a mature consideration of [women's] nobility and excellence would have proved too great a burden for his shoulders," she writes of the ancient Greek philosopher, who almost no one else dares criticize.

Marinella is not the only one who assails Passi's book. Female intellectuals line up and say what they think. Finally he seems to realize that he might be on the wrong side of history. A few years later, Passi writes a new book and titles it *The Defects of Men*.

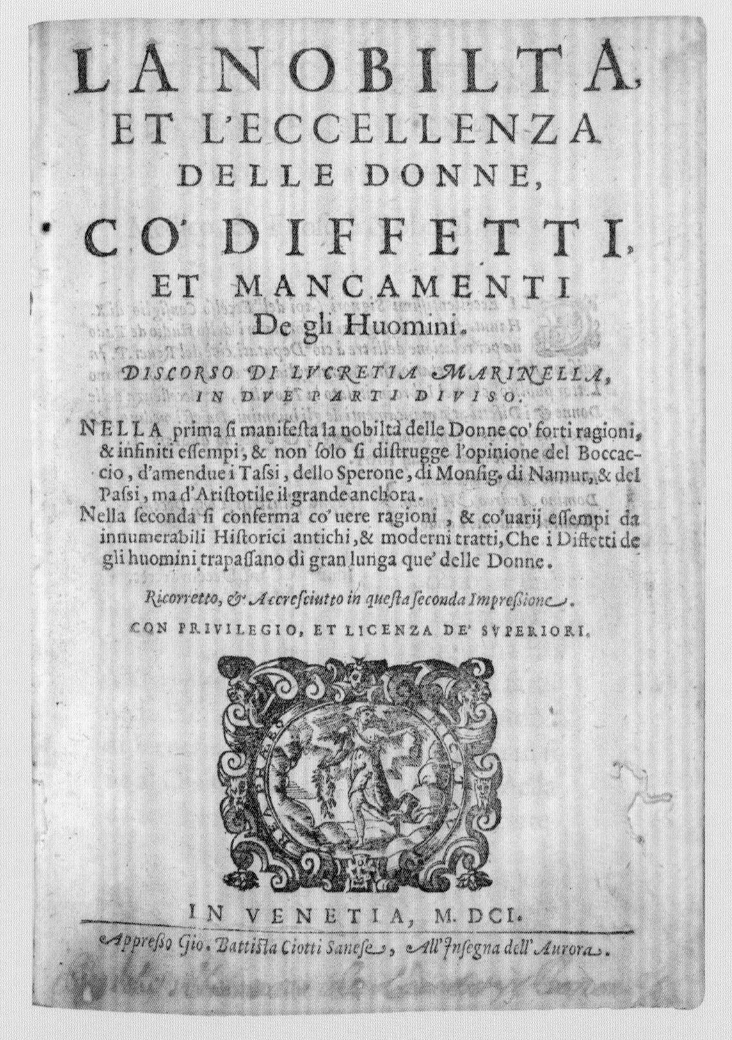

LA NOBILTA,
ET L'ECCELLENZA
DELLE DONNE,
CO' DIFFETTI,
ET MANCAMENTI
De gli Huomini.

DISCORSO DI LVCRETIA MARINELLA,
IN DVE PARTI DIVISO.

NELLA prima si manifesta la nobiltà delle Donne co' forti ragioni, & infiniti essempi, & non solo si distrugge l'opinione del Boccaccio, d'amendue i Tassi, dello Sperone, di Monsig. di Namur, & del Passi, ma d'Aristotile il grande anchora.
Nella seconda si conferma co' uere ragioni, & co'uarij essempi da innumerabili Historici antichi, & moderni tratti, Che i Diffetti de gli huomini trapassano di gran lunga que' delle Donne.

Ricorretto, & Accresciutto in questa seconda Impressione.

CON PRIVILEGIO, ET LICENZA DE' SVPERIORI.

IN VENETIA, M. DCI.

Appresso Gio. Battista Ciotti Sanese, All'Insegna dell'Aurora.

The title page of Lucrezia Marinella's scathing response to Giuseppe Passi's book The Defects of Women

created in Rome in 1801, again in 1847, once again in 1870, and yet again in 1875. In 1883, the Italian government will buy a big palace in Rome, the Corsini Palace, as a headquarters for the Lynxes, which exist today as a national academy.

Other vestiges of the Lynx legacy remain. Much of a voluminous set of natural history illustrations collected by a leading Lynx member and known as the paper museum end up being purchased by the British royal family and were recently rediscovered.

The idea of dedicated, independent scientific research, which motivated the Lynxes, will not be lost. It makes its way to the London of King Charles II, where, in 1660, it helps inspire the Royal Society. Its founding document says this of the Lynxes:

> *The members of that eminent society were exhorted to pass over in silence all political controversies, and quarrels of every kind . . . and to seek after peace, and freedom from molestation in their studies. They were required to be eager in their pursuit of real knowledge, in their study of nature and mathematics, and at the same time not to neglect the ornaments of elegant literature and philosophy.*

The Royal Society, whose home can be visited in London, was founded with the mission to "recognise, promote, and support excellence in science and to encourage the development and use of science for the benefit of humanity." It has done that very well.

The Latin motto of the Royal Society, *Nullius in verba*, can be loosely translated as "Don't take anyone's word for it." Which is very good advice for a scientist. ■

04

A Philosopher Named Bacon and a Bloody Doctor

They who have presumed to dogmatize on Nature, as on some well investigated subject, either from self-conceit or arrogance, and in the professorial style, have inflicted the greatest injury on philosophy and learning.

—Francis Bacon

I profess to learn and to teach anatomy not from books but from dissections, not from the tenets of Philosophers but from the fabric of Nature.

—William Harvey

LEFT: Francis Bacon, a royal adviser and scholar, wearing a hat and a neckpiece intended to help keep his clothes from getting soiled.

Two of the most important thinkers in Elizabethan England know each other well. One is Francis Bacon, a royal adviser, author, and intellect. The other is Bacon's doctor, William Harvey.

A statue of Bacon at the Library of Congress

As for Bacon (1561–1626), people of his time either love him or hate him. But nothing about him is boring. Ben Jonson, a playwright who knows Bacon, says of him, "There happened in my time one noble speaker. . . . His hearers could not cough or look aside from him without loss." Alexander Pope, an eighteenth-century satirist, called Bacon "the wisest, brightest, meanest of mankind."

Even today, scholars have strong feelings about Bacon. But almost everyone agrees on the greatness of his mind and the power of his writings. He is raised to be a gentleman, and despite suffering from some unfairness, and some misery he causes for himself, he remains decent and cheerful.

Bacon's father was Lord Keeper of the Great Seal of England, an important and well-paid job; he was a favorite of Queen Elizabeth I. Francis Bacon is the youngest son of the Lord Keeper's second wife. His father dies unexpectedly without having made arrangements for his children's inheritance. According to law, the oldest son inherits most of a father's wealth, so Francis doesn't inherit much. He has been brought up in sumptuous surroundings, but after his father's death, he needs to win Queen Elizabeth's favor and get a job with a big salary. Elizabeth, who has known Bacon since he was a child, recognizes his brilliance but is sometimes exasperated by him.

How the queen feels about you determines your place in London's hierarchy, and Elizabeth seems to have mixed feelings about Bacon. That may help explain why Bacon's adult life follows a yo-yo trajectory: going from misery to stardom to serious trouble to lasting acclaim.

In 1593, Bacon writes an essay criticizing a new tax, which offends the queen, who doesn't give Bacon, then a member of Parliament, a role to play in her government, depriving him of any potential income from such a post. His banishment doesn't last, though, and three years later, the queen appoints him as a counsel. The next year, in 1597, he publishes his first book. Titled *Essays*, it is given high praise by his contemporaries, and even today it holds an important place in English literature. Here is part of his essay on revenge:

> *Revenge is a kind of wild justice, which the more man's nature runs to, the more ought law to weed it out. . . . Certainly, in taking revenge, a man is but even with his enemy; but in passing it over, he is superior.*

And here is something from Bacon's essay "Of Atheism": "A little philosophy inclineth man's mind to atheism; but depth in philosophy bringeth men's minds about to religion."

Bacon is very aware of the new scientific ideas buzzing in London's intellectual world. He has a copy of Andreas Vesalius's startling book the *Fabrica*, with its carefully rendered drawings of the human body. He reads Copernicus's forbidden work on the heavens, *On the Revolution of Celestial Spheres*, and he corresponds with Galileo. He also begins writing *Instauratio Magna*, a book meant to summarize all human knowledge and establish a new kind of learning. He says scholarship should be based on observation rather than

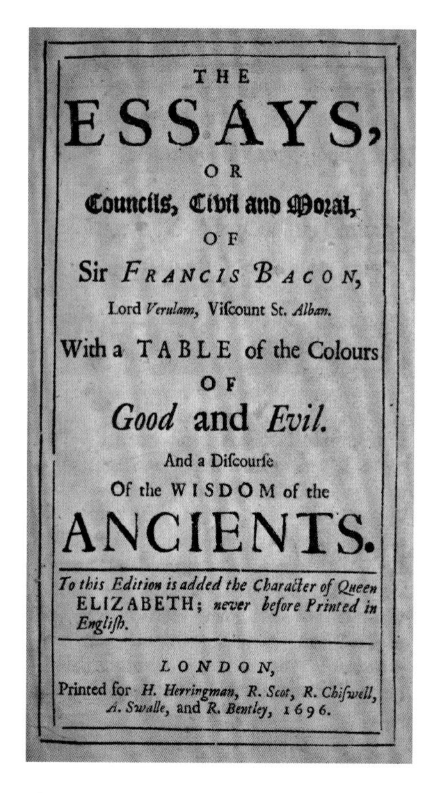
THE

ESSAYS,

OR

Councils, Civil and Moral,

OF

Sir FRANCIS BACON,

Lord *Verulam*, Viſcount St. *Alban*.

With a TABLE of the Colours

OF

Good and *Evil*.

And a Diſcourſe

Of the WISDOM of the

ANCIENTS.

To this Edition is added the Character of Queen ELIZABETH; *never before Printed in Engliſh.*

LONDON,

Printed for *H. Herringman*, *R. Scot*, *R. Chiſwell*, *A. Swalle*, and *R. Bentley*, 1696.

The title page of a 1696 edition of Bacon's Essays. *"Some books are to be tasted, others to be swallowed, and some few to be chewed and digested," Bacon writes in number 50 of his* Essays, *a surprising book that is still worth chewing and digesting.*

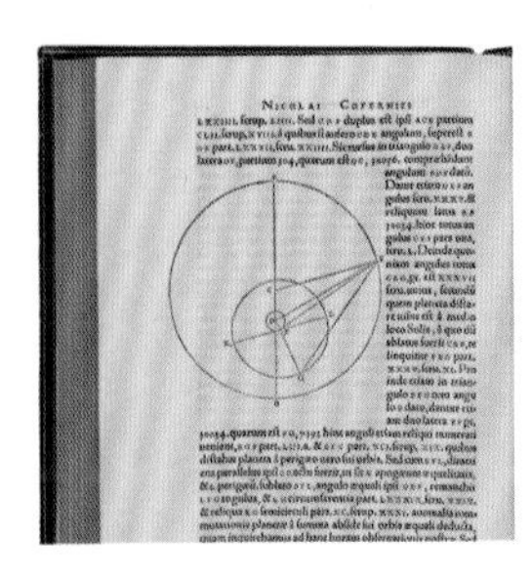

NICOLAI COPERNICI

A page of geometry from Copernicus's famous book about the movement of the planets

on ancient writings. He is laying foundations for what we now call modern science.

Meanwhile, in 1593, shy William Harvey (1578–1657) begins his medical studies. He is fifteen and enrolled at the University of Cambridge, where he will study for six years. From there he goes, in 1599, to the University of Padua, in Italy, where Galileo is a celebrity professor who teaches mathematics and astrology. Galileo has been observing the world around him and then tying observations and experiments to measurements and proofs, which help make him a successful and popular professor.

William Harvey, in 1628, explaining the circulation of the blood to King Charles I and members of his court. This 1851 engraving is from a painting by the British artist Robert Hannah.

But Harvey has come to Padua for its medical school, and there Hieronymus Fabricius (1537–1619), who teaches anatomy, is the star. Fabricius has designed a lecture hall where students can watch dissections as they listen to the professor's commentary.

Fabricius has discovered one-way valves in veins; he says they regulate the rate of blood flow in the body. William Harvey will figure out that Fabricius is wrong; it is the direction of flow that the valves control, not the rate. (The heartbeat itself is the sound of valves in the heart closing.) Harvey will also discover that blood flows continuously in the body and that it flows in only one direction.

A portrait of William Harvey with a sketch of the body's circulation system

When William Harvey returns home to England in 1602, he continues to study the human body

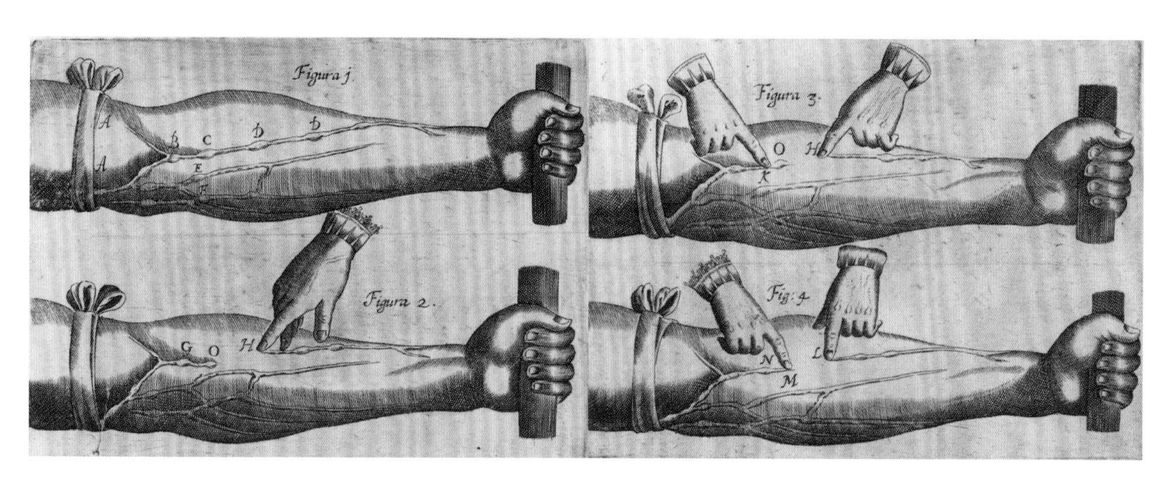

These illustrations from William Harvey's De Motu Cordis *show a demonstration of the circulation of blood, using a tourniquet to swell the veins and make them more visible.*

as few have done before. A small man with curly black hair and vibrant eyes, he marries Elizabeth Browne, the daughter of one of Queen Elizabeth's physicians. As for Queen Elizabeth, she is not immortal. After a forty-four-year reign, she dies. Her Scottish cousin James becomes the new monarch in 1603 (the King James Bible and Virginia's Jamestown will be named for him). Both Bacon and Harvey will flourish during James's rule.

In 1605, Bacon publishes *The Advancement of Learning*, which is often described as the first major philosophical book in the English language. King James I is awed by Bacon's talents, and he seems to like him too. He knights him, making him Sir Francis Bacon, and gives him important jobs, including the one Bacon wants most: the post his father held, Lord Keeper of the Royal Seal. Then the king goes further and makes Bacon Lord Chancellor of England. Bacon now becomes enormously rich. And he continues to write.

Harvey becomes an increasingly respected physician. In 1615, at age thirty-eight, he begins a series of popular lectures on the body, where he cuts into cadavers in front of an audience. The lectures to the Royal College of Physicians gain wider acclaim and make him one of London's best-known doctors. He sometimes passes around body parts and tries to educate nonscientists in the ways of medicine. Each lecture is devoted to a different area of the body: the first is the abdomen; the second, the chest; the third, the brain; and so on. What role does the heart play in the body? Dr. Harvey will begin to answer that question.

By 1618, King James names Harvey as his personal physician. Harvey also practices medicine at a London public hospital and cares for privileged private patients aside from the king. Among them is Britain's Lord Chancellor, Sir Francis Bacon.

In 1620, the same year a group of Pilgrims sails across the Atlantic to what will become known as Plymouth Colony, Bacon publishes a masterwork: *Novum Organum*. With this book, he hopes to establish a scientific process based on documented observation. To make sure that no one misses his intent, the book's title page shows a ship

An Atlantic Exchange

Eat a banana in America or a taco in Europe, and you can thank what is known as the Columbian Exchange. By the seventeenth century, Europeans are regularly voyaging across the Atlantic, and cultures that were once alien to one another are exchanging plants, animals, and diseases. The Columbian Exchange is named after Christopher Columbus by the historian Alfred W. Crosby.

Europeans bring a number of diseases to the Americas: smallpox, influenza, typhus, measles, and malaria. They also see much that was previously unknown to them, from bison to raccoons to wild turkeys. Meanwhile, many Indigenous people are curious about the horses the Spaniards ride and the wheat they use to make bread.

An Algonquian village

Soon American plants are on their way to Europe (and much of the rest of the world)—beans, cocoa, corn, peanuts, peppers, pineapples, potatoes, pumpkins, sweet potatoes, tomatoes, and vanilla. Tobacco, an American crop, will become wildly popular around the world, long before its ill health effects are known. A few animals, including turkeys and guinea pigs, will be brought to Europe as well.

A wild turkey as painted by John James Audubon

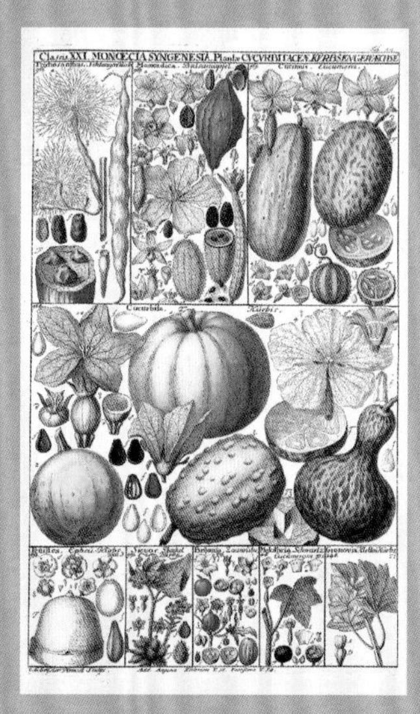

A variety of squashes native to the Americas

From Europe come many other crops headed to America, often via Africa and Asia: bananas, cattle, citrus fruits, grains, grapes, honey (and honeybees), horses, onions, peaches, pears, pigs, sheep, sugarcane, and turnips.

Today all this food moves in a circle. Coffee, which originated in Africa, was brought over the Atlantic by Europeans. Today more than half of the world's coffee is produced in Latin America. Much of it heads back over the Atlantic.

Coffee beans

sailing between the Pillars of Hercules, the ancient name for two promontories that stand at the entrance to the Strait of Gibraltar. That strait has been a boundary between the well-known Mediterranean Sea and the mostly unknown great ocean beyond. To make his point even more clear—that he is sailing into unexplored intellectual territory—he places a line from the Bible across the bottom of the title page: "*Multi pertransibunt et augebitur scientia.*" It means: "Many will travel and knowledge will be increased" (from Daniel 12:4).

The title page of Bacon's Novum Organum

Novum Organum is meant to challenge Aristotle's famous *Organon*, a work focused on grammar, rhetoric, and logic. Aristotle and his followers searched for enlightenment and truth through deep thinking. Like most of the Greeks, and most of the people of Bacon's time, they believed in the power of pure thought: if you have good minds engaging with a subject, eventually the truth will appear.

Bacon challenges that concept; for him, thinking is not enough. If you are considering science, observation and proof are essential. Bacon writes:

> *Men have sought to make a world from their own conception and to draw from their own minds all the material which they employed, but if instead of doing so, they had consulted experience and observation, they would have the facts and not opinions to reason about, and might have ultimately arrived at the knowledge of the laws which govern the material world.*

Laws govern the material world? Almost no one believes that. Most people think it is God's direct action that creates change. Since God's decisions are impossible to predict, how can there be laws of nature? Bacon and a few others of his time are considering that dilemma. Are they out of touch with reality or ahead of their time?

As it turns out, they are ahead of their time. Thinkers to come will agree: there are laws of nature. Bacon's approach, today's modern way of doing science, depends on proof. The universe does have laws, or rules, and they can be tested.

Understanding those laws will become a goal of modern science. Finding them will take more than imagination: it will demand experimentation and tests and proofs.

Bacon's big ideas, that science can't just come out of someone's head and that science's concepts must be provable, are a step toward a new kind of scientific thinking. Bacon is often given credit for founding the scientific method; that isn't quite true, but he does take a significant step in that direction. He makes some big mistakes too, one of which is to reject Copernicus's great idea that the sun lies at the center of the solar system. But he can be credited with the shift toward the search for scientific proof.

Today we define the scientific method as a process in which a problem is identified, relevant observations and experiments lead to a hypothesis, then that hypothesis is tested, often through experiments and observations that control for as many variables as possible. If the hypothesis passes the experimental test, you have scientific validation; given enough passed tests, you have a scientific law, which holds unless someone disproves it.

Bacon will be best remembered as an author, but historians will debate his role as an adviser to the king. In that role, he is wise and discerning but also controversial. When the Earl of Essex is accused of treason after bringing a small army into London, it is Bacon who helps preside over his friend's conviction and execution. Some of London's elite will never forgive him.

When it comes to his own affairs, Bacon spends money on grand homes, fancy clothes, and legions of servants. He is accused of financial corruption and admits to taking bribes from people who want favors. Some modern scholars believe that his failings were greatly distorted by his rivals, but Bacon is briefly sent to jail in the Tower of London. He handles that public humiliation with dignity. After that, Bacon is barred from public service and mostly focuses on his writings and their world-changing ideas.

Bacon mentions Gaius Plinius Secundus, known as Pliny the Elder (23–79 CE), in his writings, and everyone who can read (and many who can't) in the England of Queen Elizabeth, King James, and Shakespeare knows about him as well. He was a scientist as well as a Roman military leader. He was also a writer who wrote nearly a hundred books. Have you ever heard the term "in a nutshell"? That comes from Pliny, who once wrote that he heard about someone who had written down a version of the *Iliad* that was so small it could fit in a nutshell.

Pliny's most important book was called *Naturalis historia* (*Natural History*) and was sort of an encyclopedia of all the world's knowledge of life science. He also pioneered the idea of verified scientific observation, which makes him a grandparent of modern science and put him way ahead of many who followed him. Pliny died trying to rescue a friend when Mount Vesuvius erupted and lava destroyed the town of Pompeii.

TOP LEFT: A statue of Pliny the Elder at a cathedral in Como, Italy; BOTTOM LEFT: A 1750 etching depicting the eruption of Mount Vesuvius in 79 CE

He writes a novel, *The New Atlantis*, describing a great (fictional) research university where nature is studied directly; that innovative idea will become influential in education circles to come. But in 1627, when the book is published, Bacon is not around for the event.

A year earlier, in 1626, he attempts an experiment: he tries to freeze a duck in the snow to observe if cold preserves meat. Bacon gets an illness that leads to pneumonia. He is carried to a nearby mansion belonging to Lord Arundel.

As the king's physician, William Harvey is called to examine four women accused of witchcraft. It is 1634, and belief in witches is widespread, and so are witch trials. Harvey, who has a scientific mind, looks for evidence or proof of witchcraft; he discovers none and finds the alleged witches innocent.

A scene with witches and demons from a 1693 book published in Germany

Realizing that he is not going to recover, Bacon writes a letter about his impending demise to Arundel, describing himself as a martyr to the scientific method:

> *My very good Lord—I was likely to have had the fortune of Caius Pliny the Elder, who lost his life by trying an experiment about the burning of Mount Vesuvius; for I was also desirous to try an experiment or two touching the conservation and induration of bodies. As for the experiment itself, it succeeded excellently well.*

As for Bacon, no one can save him. Not even his brilliant physician, Dr. William Harvey. Two years after Bacon's frigid departure, Harvey publishes his masterwork, *Exercitatio Anatomica de Motu Cordis et Sanguinis in Animalibus* (*On the Motion of the Heart and Blood in Animals*), commonly known as *De Motu Cordis*. In it he says that the heart pumps blood around the body and then recirculates it, which means the blood in the arteries is the same blood as that in the veins. It's a closed system. Harvey's discovery of the circulatory system is a big moment in medical history and in life science too.

Most doctors of the time do not accept Harvey's theory of blood circulation. They believe the great Roman physician Galen, who wrote that blood in the veins is totally separate from that in the arteries and that the two never mix. Doctors are taught, and most incorrectly believe, that food turns into blood in the liver and then becomes fuel for the body, with the liver constantly creating new blood.

With the medical establishment against him, Harvey loses patients. He retires to a living room filled with books. When a friend tries to lure him back into active practice, he writes, "Much better is it oftentimes to grow wise at home and in private, than by publishing what you have amassed . . . to stir up tempests that may rob you of peace and quiet for the rest of your days."

When Harvey writes about Bacon, he says, "He writes philosophy [meaning science] like a lord chancellor," which is not meant as a compliment. But Bacon's writings will help change the way we consider the whole living world; Harvey's will change our understanding of blood flow. Each of them will have a profound effect on the field of life science that is being born. ■

ART OF
ERRANEAN
A.
ALPS
GENOA
GULF OF GENOA
LEGHORN
TUSCANY
ITALY
ISLE OF CORSICA
STRAITS OF BONIFACIO
ISLE OF SARDINIA
ROME
NAPLES
Gulf of Naples
CALABRIA
GULF OF TARENTO
SICILY
MESSINA
Palermo
Catania
CHANNEL OF MALTA
Malta
Stromboli
Lipari
DALMATIA
GULF OF VENICE
VENICE
Ancona
VARIATION 17° WEST.
Gulf of Manfredonia
Manfredonia
Bari
Brindisi
Tarento
Otranto
Corfu
Narento
Ragusa
Cataro
Durazzo
Valona
FRIKIA
ARY
KINGDOM OF TUNIS
TUNIS
TRIPOLI
GULF OF SIDRA
Mountains of Benoletta
Mount Negro
EGHORN
THIS

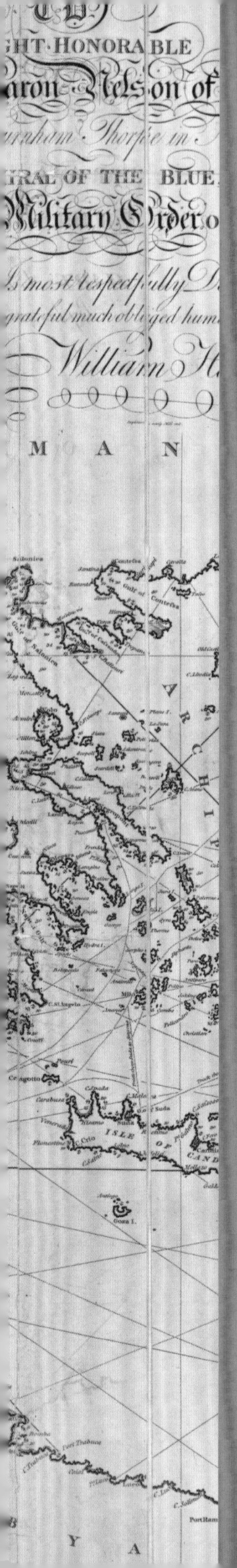

05

Tongues That Are Teeth: A Shark, Steno, and the Cimento

If by some fiat I had to restrict all this writing to one sentence, this is the one I would choose: The summit of Mount Everest is marine limestone.

—John McPhee

Should philosophy guide experiments, or should experiments guide philosophy?

—Liu Cixin

LEFT: A 1797 map of the Mediterranean Sea

In 1666, some fishermen at Livorno, on Italy's west coast, catch a huge, live female white shark. The frightening fish, with rows of sharp pointed teeth and with roe in her innards, weighs more than a ton.

In his palace in Florence, the Grand Duke of Tuscany, Ferdinando II, learns of the big catch and, being scientifically inclined, asks for the shark's head. He wants to see what can be learned by studying it.

So the head is chopped off, loaded on a cart, and sent his way. Since the body is beginning to rot and shark isn't thought edible, the rest of the cadaver is tossed back into the sea.

Ferdinando II. Unfortunately, his feathers and fancy clothes didn't give him leadership skills.

Ferdinando II (1610–1670) is a Medici, part of a famous family that has long dominated politics and the arts in vibrant Florence. He was ten years old in 1621, when his commanding father died, which made the boy a grand duke; for a while his grandmother served as regent. He was twenty-three when his tutor, Galileo, was hauled before the Inquisition; the young duke didn't have the political savvy to protect the stargazer as his powerful father might have done.

It is two decades after Galileo's humiliation, and Ferdinando and his younger brother Leopoldo now control the money and power that come with their place in Italy at that time. They use part of their inheritance to finance a think tank of natural philosophers; its members are followers of the now-dead Galileo. Called the Accademia del Cimento (Academy of Experiment), it is heir to the pioneering Academy of the Lynxes. The Cimento members are determined to make observation and experimentation central to their research.

The Cimento scholars and their patrons in Florence are eager to look inside the skull of the big fish. And they know who should do the cutting.

Nicholas Steno (1638–1686), a twenty-eight-year-old Dane named Niels Stensen at birth, has Latinized his name and come to Florence because of the reputation of the Cimento. More than a hundred years have gone by since the Paduan professor Vesalius published his revolutionary book, the *Fabrica*, with its beautifully rendered pictures of bodies and their parts. Dissections are now performed regularly at centers of sophistication in Europe. Princes and kings and those who can pay for a seat watch as corpses are sliced; the surgeon usually provides a running commentary and passes around body parts so those who attend can get a close look.

Steno is *the* superstar in this field. The year before his arrival in Tuscany, a doctor visiting Paris writes, "M. Steno is the rage here. This

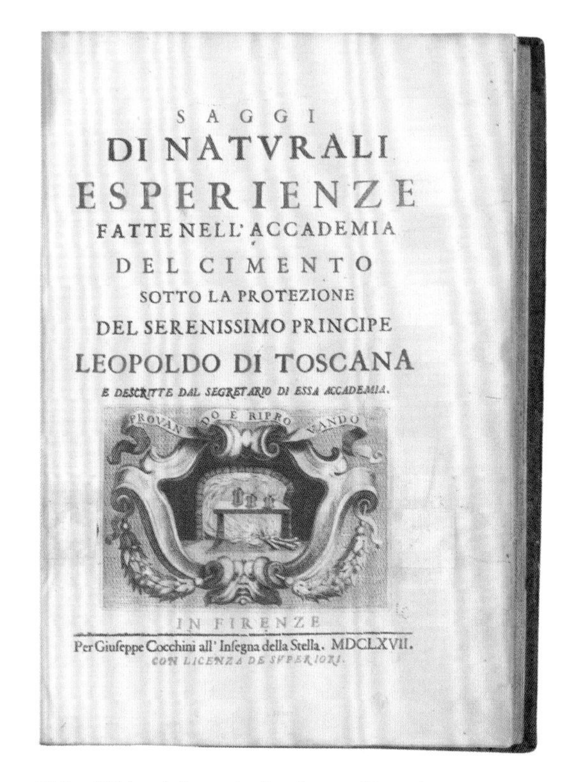

SAGGI
DI NATVRALI
ESPERIENZE
FATTE NELL' ACCADEMIA
DEL CIMENTO
SOTTO LA PROTEZIONE
DEL SERENISSIMO PRINCIPE
LEOPOLDO DI TOSCANA
E DESCRITTE DAL SEGRETARIO DI ESSA ACCADEMIA.
PROVANDO E RIPROVANDO
IN FIRENZE
Per Giuſeppe Cocchini all' Inſegna della Stella. MDCLXVII.
CON LICENZA DE SVPERIORI.

This 1666 book from the Academy of Experiment was dedicated to Ferdinando II.

The year the shark is brought ashore, 1666, is a big one in science. That very same year, a juicy apple falls in Isaac Newton's mother's garden; her son connects the fall of the apple to the orbit of the moon around Earth, which leads him to thoughts about gravitation. *Why do objects fall down instead of up?* he asks himself. His answer will change the world.

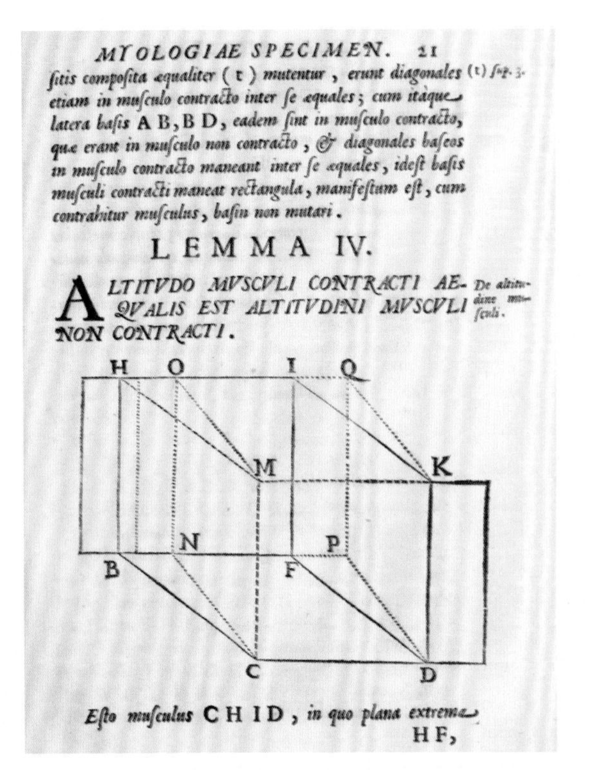
MYOLOGIAE SPECIMEN. 21

ſitis compoſita æqualiter (t) mutentur, erunt diagonales etiam in muſculo contracto inter ſe æquales; cum itàque latera baſis A B, B D, eadem ſint in muſculo contracto, quæ erant in muſculo non contracto, & diagonales baſeos in muſculo contracto maneant inter ſe æquales, ideſt baſis muſculi contracti maneat rectangula, manifeſtum eſt, cum contrahitur muſculus, baſin non mutari.

(t) ſup. 3.

LEMMA IV.

ALTITVDO MVSCVLI CONTRACTI AEQVALIS EST ALTITVDINI MVSCVLI NON CONTRACTI.

De altitudine muſculi.

Eſto muſculus C H I D, in quo plana extrema HF,

A bit of geometry from a 1667 book by Nicholas Steno

evening after dinner we saw him dissect the eye of a horse . . . we are only apprentices next to him."

Cutting into a human skull, Steno finds previously unknown tear ducts. Using geometry, Steno shows that as a muscle contracts, it changes shape but not volume. Later, during another dissection, he discovers eggs in a female ovary. Before that, eggs were thought to occur only in chickens and other egg-laying creatures, not in species like humans that give birth to live offspring. Steno's dissections make it clear that females are more than carriers of male-generated embryos, as has been generally believed.

So, of course, Steno is the one to carve up the shark. Duke Ferdinando II, who can afford the best, expects Steno not only to dissect the fish's head but also to write a scholarly paper about what he finds, which he does. Cutting into the sea creature, Steno is surprised by the tiny size of its brain. How can such a small brain control such a large animal? None of the current theories on brain function can explain it. As for the mouth and jaws, Steno says they are big enough to "let a whole man pass through them without difficulty." But it is the shark's teeth that especially fascinate the young Dane. The shark has several rows of teeth, and the ones in the outer row are sharp triangular blades, serrated for slicing and dicing.

Steno knows that triangular stones, called tongue stones, can be found strewn about on Mediterranean beaches, especially on the island of Malta. The stones are said to have curative powers. Some men hang them around their neck and claim they attract beautiful women.

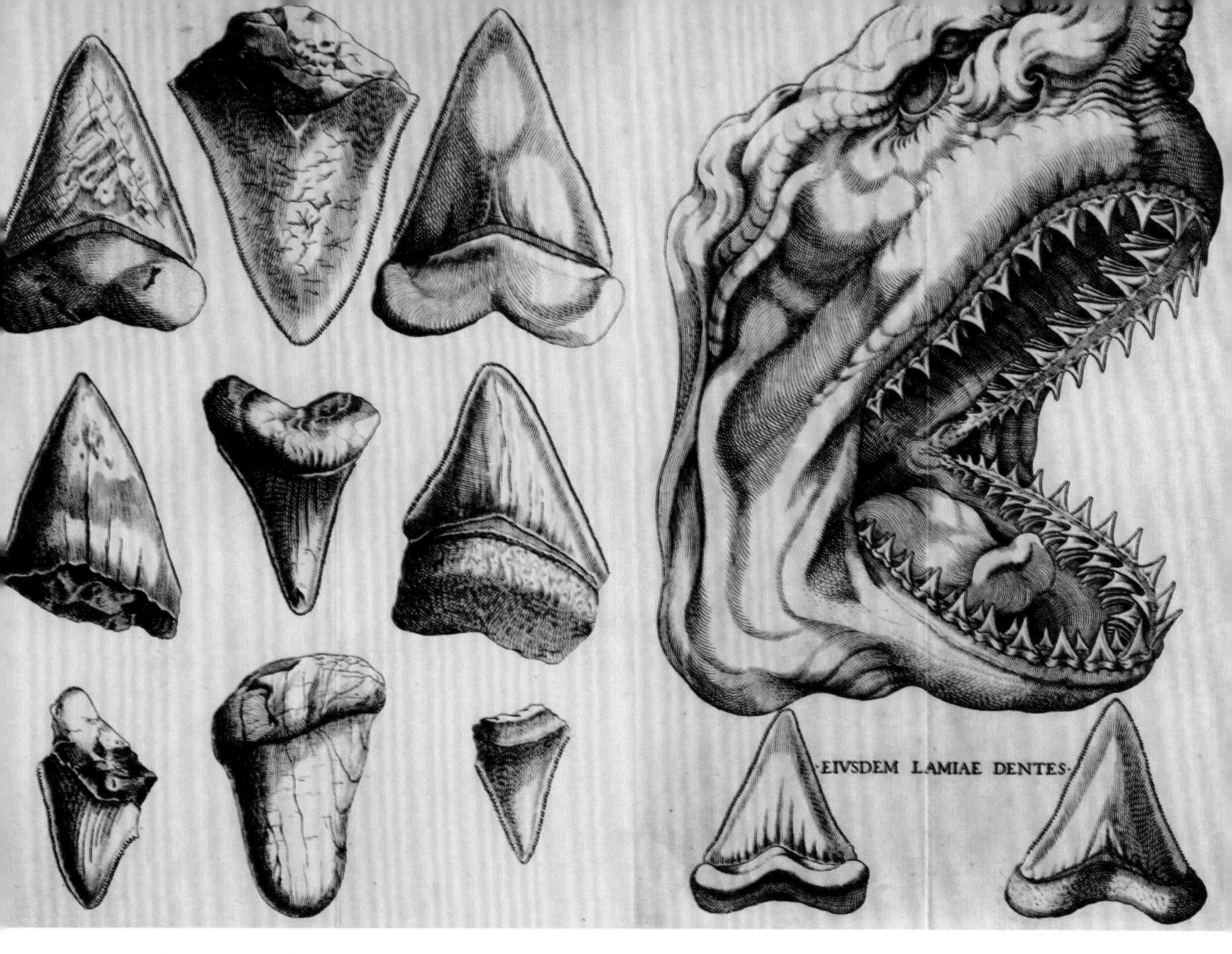

Steno's illustrations of the shark's head obtained by Ferdinando II and of its teeth, which particularly intrigued him

Tongue stones are also commonplace on Vatican Hill, in Rome, home to the popes. No one has been able to figure out what they are. Steno, studying those stones after he has dissected the shark, suggests they may be fossilized sharks' teeth.

There is a bigger issue here than just tongue stones. Why are they on a hill? If they are sharks' teeth, how did they get there, so far from a beach? What about seashells found on mountaintops? If plants can grow out of the earth, could shells and tongue stones grow too?

Sharks have at least five rows of teeth, which they continually shed. Some shed as many as thirty-five thousand teeth in a lifetime.

TABLEAU CHRONOLOGIQUE

des divers terrains ou systèmes de couches minérales stratifiées qui constituent la partie connue de l'écorce terrestre ; présentant d'une manière synoptique les principaux êtres organisés qui ont vécu aux diverses époques géologiques, et indiquant l'âge relatif des différents systèmes de soulèvements de montagnes établis par Mr. Elie de Beaumont.

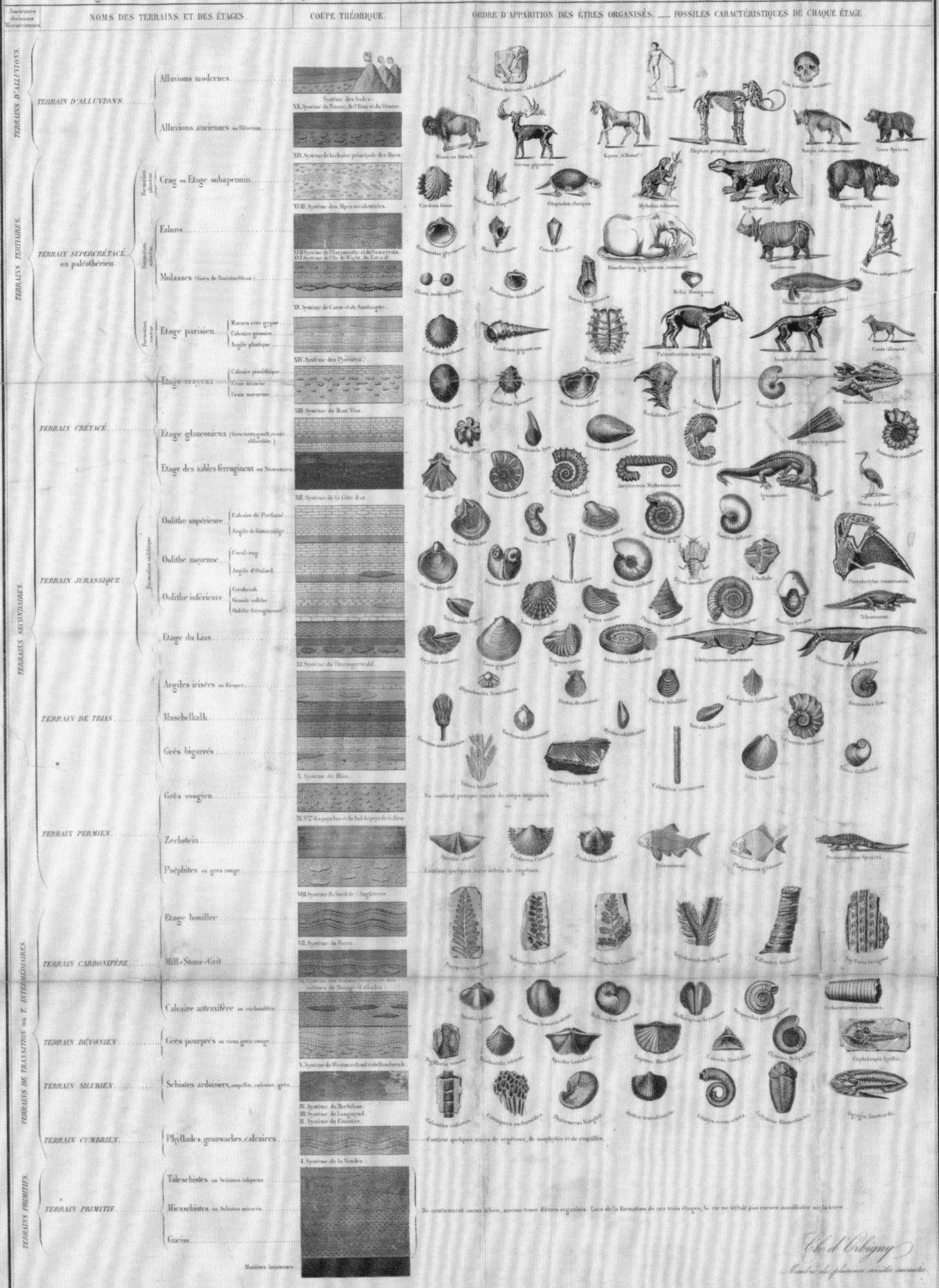

Anciennes divisions Wernériennes	NOMS DES TERRAINS ET DES ÉTAGES.	COUPE THÉORIQUE.	ORDRE D'APPARITION DES ÊTRES ORGANISÉS. — FOSSILES CARACTÉRISTIQUES DE CHAQUE ÉTAGE.

TERRAINS D'ALLUVIONS.

- *TERRAIN D'ALLUVIONS.*
  - Alluvions modernes.
    - Système des Andes.
    - XX. Système du Ténare, de l'Etna et du Vésuve.
  - Alluvions anciennes ou Diluvium.
    - XIX. Système de la chaîne principale des Alpes.

TERRAINS TERTIAIRES.

- *TERRAIN SUPERCRÉTACÉ ou paléothérien.*
  - Formation pliocène
    - Crag ou Etage subapennin.
      - XVIII. Système des Alpes occidentales.
  - Formation miocène
    - Faluns.
      - XVII. Système de l'Erymanthe et du Sancerrois.
      - XVI. Système de l'île de Wight, du Tatra &c.
    - Molasses (Grès de fontainebleau).
      - XV. Système de Corse et de Sardaigne.
  - Formation éocène
    - Etage parisien.
      - Marnes avec gypse
      - Calcaire grossier
      - Argile plastique
      - XIV. Système des Pyrénées.

TERRAINS SECONDAIRES.

- *TERRAIN CRÉTACÉ.*
  - Etage crayeux.
    - Calcaire pisolithique
    - Craie blanche
    - Craie marneuse
    - XIII. Système du Mont Viso.
  - Etage glauconieux (Grès vert, gault, craie chloritée.)
  - Etage des sables ferrugineux ou Néocomien.
    - XII. Système de la Côte d'or.
- *TERRAIN JURASSIQUE.*
  - Formation oolithique
    - Oolithe supérieure
      - Calcaire de Portland
      - Argile de Kimmeridge
    - Oolithe moyenne.
      - Coral-rag
      - Argile d'Oxford
    - Oolithe inférieure
      - Cornbrash
      - Grande oolithe
      - Oolithe ferrugineuse
  - Etage du Lias.
    - XI. Système du Thuringerwald.
- *TERRAIN DE TRIAS.*
  - Argiles irisées ou Keuper.
  - Muschelkalk.
  - Grès bigarrés.
    - X. Système du Rhin.
- *TERRAIN PERMIEN.*
  - Grès vosgien.
    - IX. Systes. des pays-bas et du Sud du pays de Galles.
  - Zechstein.
  - Psëphites ou grès rouge.
    - Contient quelques rares débris de végétaux.
    - VIII. Système du Nord de l'Angleterre.

TERRAINS DE TRANSITION ou T. INTERMÉDIAIRES.

- *TERRAIN CARBONIFÈRE.*
  - Etage houiller.
    - VII. Système du Forez.
  - Mill-Stone-Grit.
  - Calcaire antraxifère ou carbonifère.
- *TERRAIN DÉVONIEN.*
  - Grès pourprés ou vieux grès rouge.
    - V. Système du Westmoreland et du Hundsrück.
- *TERRAIN SILURIEN.*
  - Schistes ardoisiers, ampélite, calcaire, grès.
    - IV. Système du Morbihan.
    - III. Système du Longmynd.
    - II. Système du Finistère.
- *TERRAIN CUMBRIEN.*
  - Phyllades, grauwackes, calcaires.
    - Contient quelques traces de végétaux, de zoophytes et de coquilles.
    - I. Système de la Vendée.

TERRAINS PRIMITIFS.

- *TERRAIN PRIMITIF.*
  - Talcschistes ou Schistes talqueux.
  - Micaschistes ou Schistes micacés.
  - Gneiss.
  - Ne contiennent aucun débris, aucune trace d'êtres organisés. Lors de la formation de ces trois étages, la vie ne s'était pas encore manifestée sur la terre.
  - Matières ignées.

Ch. d'Orbigny
Membre de plusieurs sociétés savantes.

Nota : *Les terrains d'épanchement (granitoïde, porphyroïde), et d'éruption (trachyto-basaltique et lavique), ne figurent pas sur ce tableau.*

F. SAVY, Libraire de la Société Géologique de France, 24 Rue Hautefeuille, Paris.

LEFT: This 1850 chart from France shows fossils found at different layers of rock, with older fossils found at lower layers.

Leonardo da Vinci had an answer. Observing floods in northern Italy nearly two centuries earlier, he wrote in his notebook, "The stratified stones of the mountains are all layers of clay, deposited one above the other by the various floods of the rivers." He figured out that floods and other natural activities change the Earth, have done so through time, and often leave layers of stone that document the passage of time.

As far as we know, Steno never sees Leonardo's writings. He has to rediscover this seminal idea: Earth's history is a tale of change, and those changes are recorded in Earth's layers. That idea challenges the widely accepted belief that Earth is exactly as it was at its creation.

Believing that all of Earth's early rocks and minerals were once fluid (he is right), Steno reasons that as this fluid hardens, particles and sediment fall to its bottom, leaving a horizontal layer. Another layer may form on top of that first layer. And another. Which means that the bottom layers in rock strata are the oldest; it also means that layered rocks on mountainsides are layers of time. Steno's concept, known as the law of superposition, is basic to today's geology, and he will later be considered a founder of geology.

So how do fish fossils find their way to mountaintops? Steno realizes that some mountains have risen from the sea through natural processes of decay and building that can take much time. Today we know that means millions or even billions of years.

Steno's big idea that the fossil record is a chronology of living creatures laid down over eons will help support two theories that are incubating. One is that the Earth is very old, much older than previously imagined, and that it has changed over time. The second is that life-forms have also changed and had a great expanse of time to do it. Those ideas, based on the new scientific thinking, will have to compete with the old familiar concept that has been taught in schools and churches for generations: that Earth and life today are much as they were at the Creation. ■

06

Spontaneous? Why Not?

A scientist in his laboratory is not only a technician: he is also a child placed before natural phenomena which impress him like a fairy tale.

—Marie Curie

The senses are scouts, or spies, that seek to discover the nature of things, and report these observations to Reason within.

—Francesco Redi

LEFT: These little figures on the Grand Staircase at the Library of Congress represent an electrician and an astronomer.

Francesco Redi (1626–1697), born in Arezzo, Italy, is a poet, natural philosopher, literary scholar, stargazer, and Cimento member. He is also personal physician to Duke Ferdinando II. He writes notes on the heavens that even today, almost four centuries later, are still pertinent. Which may be why, in 2014, an impact crater on Mars is named Redi.

Like many intellectuals in his time, as in ours, Redi is concerned with the how-did-life-begin question. Aristotle made notes as he watched maggots, beetles, and flies emerge out of rot; it seemed clear to him and others that some life-forms arise from decay. That explanation leads to a concept known as spontaneous generation. To those using observation to inform their thinking, spontaneous generation makes sense.

Prior to Redi, Jan Baptista van Helmont (1579–1644), a Flemish doctor and chemist, wrote a recipe for creating eels: "Cut two pieces of grass sod wet with Maydew and place

It is Jan Baptista van Helmont who introduces the concept of gas, and the word for it, to the scientific world. He also does a famous experiment with a willow tree. In his time, people believe that plants grow by eating soil. Van Helmont decides to check that. He plants a willow tree in a pot of dry soil, which he waters for five years. Then he weighs the tree. It has increased in weight by 168 pounds. The pot of soil weighs about the same as it had. He concludes that the increase in the mass of a plant comes from water. He is close: it actually comes from water *and* carbon dioxide in the air.

The title page of Van Helmont's 1682 book Opera Omnia *shows him and his son along with several coats of arms. The lower image depicts an alchemist and a mine where men are working.*

LEFT: An engraved portrait of Francesco Redi with fashionably long curls and a serious gaze

the grassy sides together, then put it into the rays of the spring Sun, and after a few hours you will find that a large number of small eels have been generated."

Do you want mice? Van Helmont said to put sweaty underclothes in a jar with wheat, wait for twenty-one days, and the sweat will turn the wheat husks into mice. If you have doubts about this concept, some of those who believed it would say at the time, just go to Egypt, where each year frogs appear spontaneously when the Nile River overruns its banks. The frogs aren't there when the banks are dry, so they must be formed out of the river's mud.

The now-esteemed London physician William Harvey doesn't buy the spontaneous generation idea; he believes that flies and other vermin arise from eggs too small to be seen. Sophisticated microscopes will eventually confirm that he is right, but in the seventeenth century, he is way ahead of most people.

The Catholic Church in Rome agrees with Harvey; it does not accept the spontaneous generation concept either. Every creature is the direct and unique creation of God, say the church fathers. Redi, who is Jesuit trained, agrees. He intends to prove the church's position. In 1668, two years after the shark incident, he sets up a test to see if insect life will form on its own in a noxious but carefully covered brew of rot.

ESPERIENZE
INTORNO ALLA GENERAZIONE
DEGL' INSETTI
FATTE
DA FRANCESCO REDI
ACCADEMICO DELLA CRUSCA,
E DA LVI SCRITTE IN VNA LETTERA
CARLO DATI
IN FIRENZE

Harvey guessed that insects do not arise spontaneously. Redi proved it, as described in his book published in 1668 in Florence.

Redi places smelly meat in boxes screened with muslin. He also puts meat in unscreened boxes. Maggots appear where there are no screens but not where screens are present. In this experiment, he shows that maggots are the larvae of flies. He has made it clear: the maggots have *not* been generated spontaneously.

In 1668, Redi's report on his experiments is published. The book is called *Esperienze intorno alla generazione degl'insetti* (*Experiments on the Generation of Insects*). It doesn't confirm Harvey's theory about flies coming from eggs, but it does strengthen the church's stand that life cannot form spontaneously. And it gives Redi a place as a founder of experimental biology.

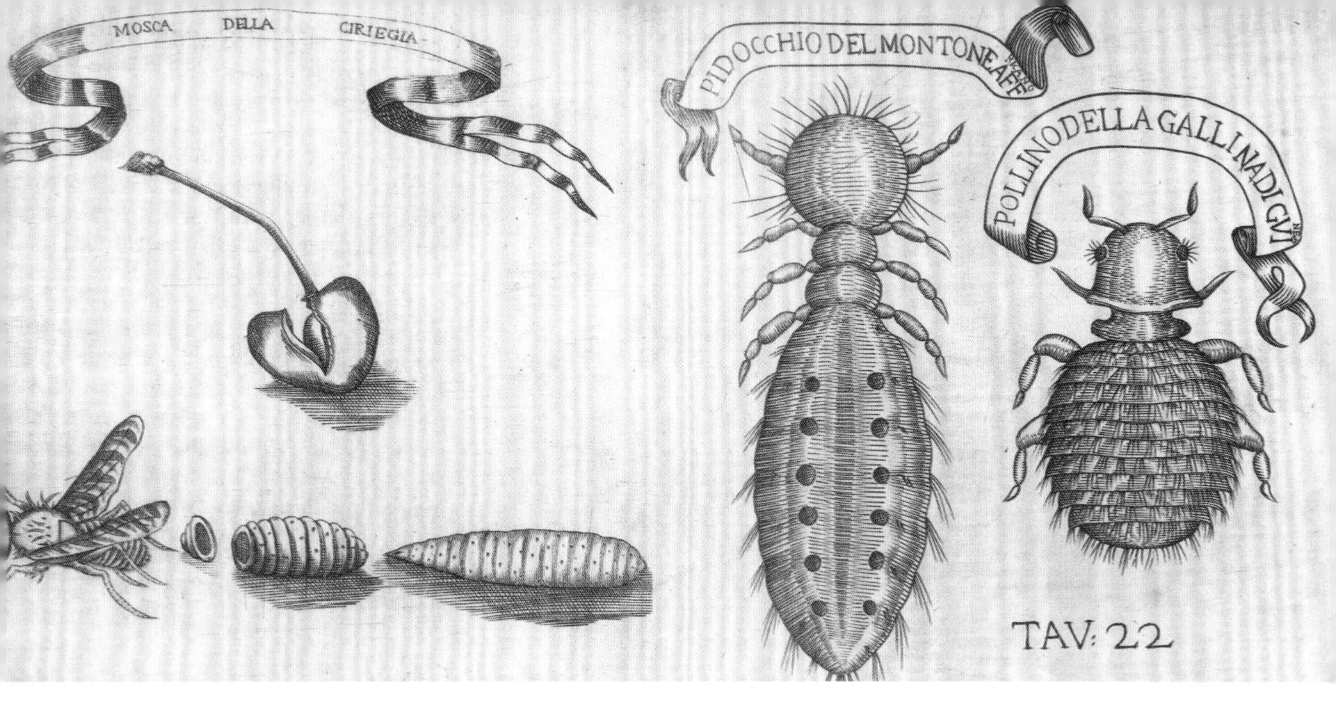

Illustrations from Redi's 1668 book show the sequence of a fly developing from a maggot found in a cherry (left), as well as two types of lice (right).

But it will take more than one experiment to kill the idea of spontaneous generation. Almost everyone has seen a rotting carcass; flies and bees do seem to appear out of them.

So the concept persists. Since it runs counter to traditional church teaching, a belief in spontaneous generation is sometimes equated with daring progressive thinking. As time passes and microscopes improve, more people see microscopic creatures wiggling about in water or in spit; those *animalcules* become central to the debate on spontaneous generation.

In Great Britain, those who hear of Redi's experiment have doubts. The English are not going to be convinced by a study done by an Italian. They need to hear those ideas from an English scientist. So, almost a century after Redi, an English Catholic priest, John Needham (1713–1781), decides to see if he can prove (or disprove) spontaneous generation.

In 1745, Needham boils mutton soup. The boiling is meant to kill any organisms in the soup. Then he puts the broth in a series of flasks and seals the flasks with cork. Weeks later, when he pulls out the corks, microorganisms are happily wiggling in all the flasks. Needham believes his experiment has settled things. It seems clear that some life-forms have been created naturally out of the rancid soup.

John Needham using a candlelit microscope to see life in new ways

A statue of Lazzaro Spallanzani in his hometown of Scandiano, Italy, depicts him studying a frog through a magnifying glass. Unveiled in 1888, the statue is the work of local artists Guglielmo Fornaciari, who made the figure, and Vasco Montecchi, who sculpted the frog.

A few decades after that, an Italian priest, Lazzaro Spallanzani (1729–1799), questions Needham's results. Spallanzani, who uses a microscope as well as anyone in his time, has spent hours looking at wiggling creatures that the naked eye can't see.

Spallanzani has been trained by his cousin Laura Bassi, the first woman in all of Europe to become a full-fledged college professor (a few women have taught at universities earlier, but with secondary status). Bassi devotes herself to mathematics and experimental physics, publishes twenty-eight scientific papers, and is active in several literary societies, in addition to raising eight children.

Born in Bologna, Italy, Laura Maria Catarina Bassi (1711–1778) amazes everyone with her knowledge, including the ability to speak Italian, French, and Latin—even as a child. The family doctor oversees her education. Even the archbishop of Bologna visits the child prodigy in his midst.

Unlike other women of her time, Bassi is allowed to enroll at the University of Bologna. She becomes the first European woman to earn a doctorate in science. She studies Newton's big ideas (including calculus) before most of her colleagues understand them. Students travel to Bologna to study with her, often at her home because of restrictions placed by her university about how much a woman can teach. She also petitions to be allowed to read science books banned by the Catholic Church.

When Pope Benedict XIV puts together a scholarly think tank called the Benedictines, he doesn't include her. Bassi complains to the pope. He relents and adds her to the elite group.

A relative of Bassi, Agostino Bassi (1773–1856), is also a player on the biological stage. His lawyer father expects him to study law, but Agostino Bassi has a passion for microscopes and science. Studying tiny parasites, he figures out that diseases, including those that infect humans like syphilis and the plague, are caused by microorganisms that are alive. This is an insight way ahead of its time.

Laura Bassi

So independent thinking is a family thing with the Bassis and certainly with Spallanzani, who decides to try to settle the spontaneous generation question once and for all. Spallanzani hypothesizes that cork, which is porous, may not keep out organisms too small to be seen with the naked eye.

He does his own experiment, copying Needham's, but after boiling the broth, he seals his flasks by melting their glass necks, thus fusing them shut. When he breaks into those flasks weeks later, there are no animalcules wiggling about; there has been no spontaneous generation. Still, some people won't accept his results. The spontaneous generation idea is hard to kill. ■

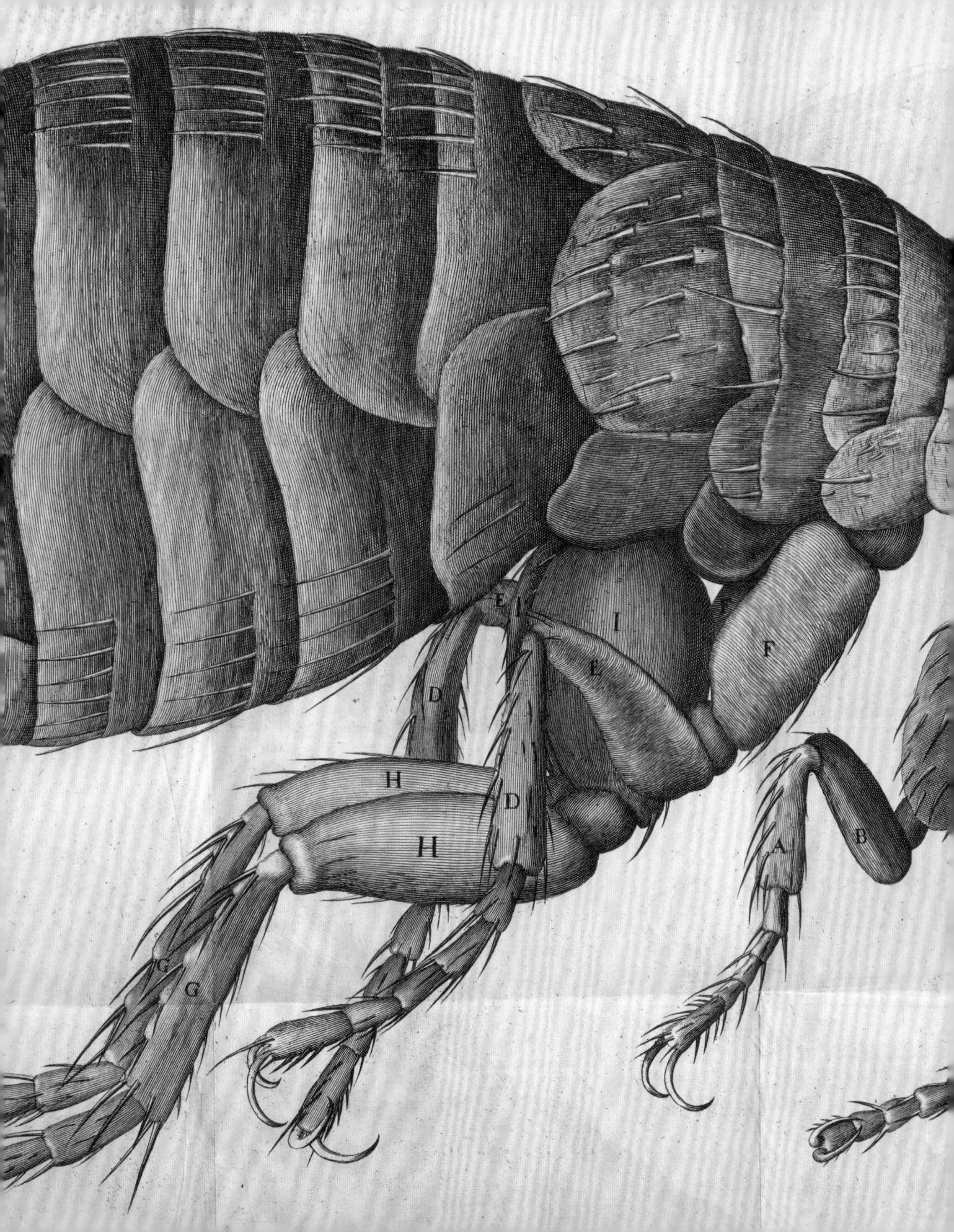
E
I
F
I
F
E
D
H
D
H
B
A
G
G

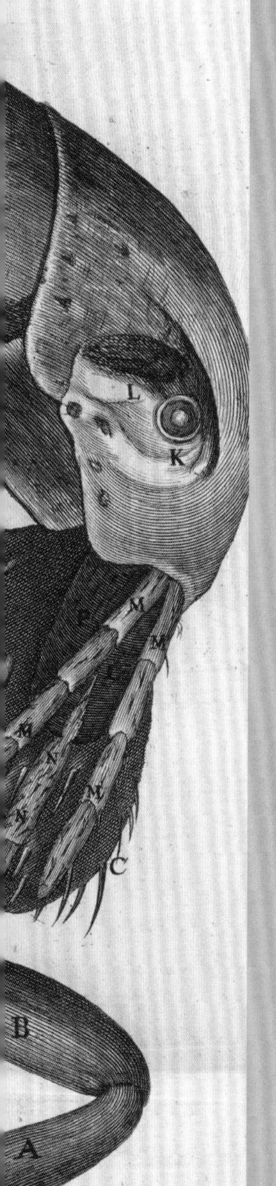

07

Magnified Wonders Help Create an Awesome Book

The next care to be taken, in respect of the Senses, is a supplying of their infirmities with Instruments, and, as it were, the adding of artificial Organs to the natural; this in one of them has been of late years accomplisht with prodigious benefit to all sorts of useful knowledge, by the invention of Optical Glasses. By the means of Telescopes, there is nothing so far distant but may be represented to our view; and by the help of Microscopes, there is nothing so small, as to escape our inquiry; hence there is a new visible World discovered to the understanding.

—Robert Hooke

The abiding fact is that modern science grew out of the lovely Medieval idea of ordo mundi, *the faith in a universal order, a religious feeling for the ultimate unity of all life.*

—Thomas Goldstein

LEFT: Close-up of a flea from Robert Hooke's Micrographia

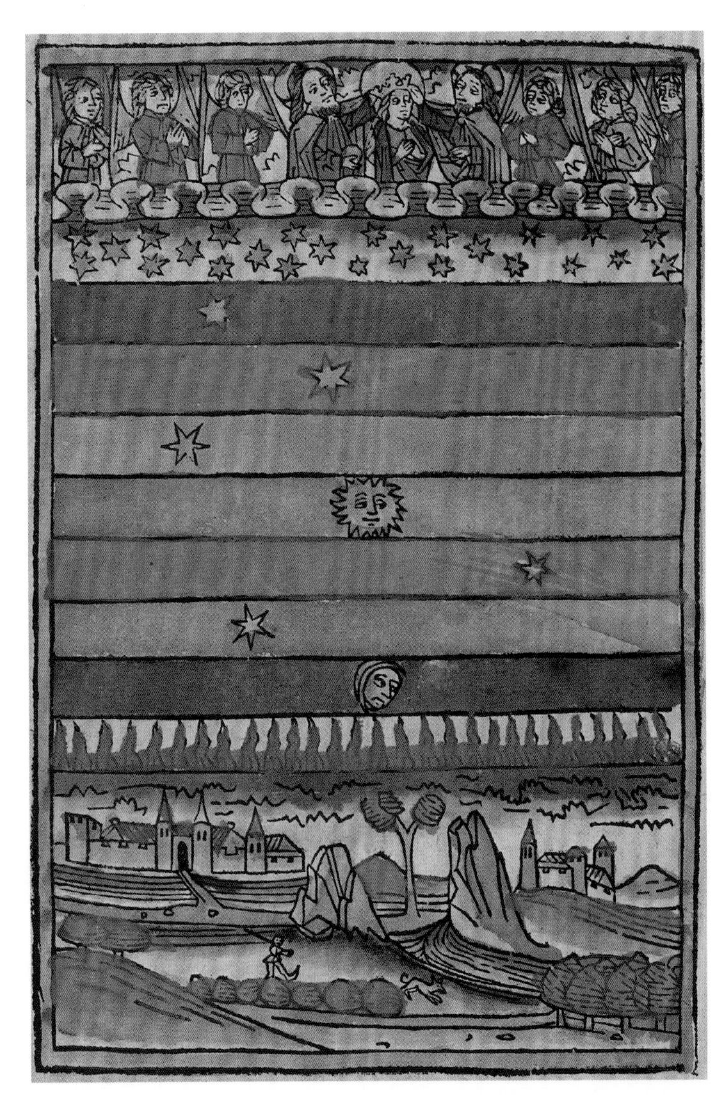

A depiction of Earth and the heavens from a medieval science compendium

MICROGRAPHIA:
OR SOME
Physiological Descriptions
OF
MINUTE BODIES
MADE BY
MAGNIFYING GLASSES.
WITH
OBSERVATIONS and INQUIRIES thereupon.

By *R. HOOKE*, Fellow of the ROYAL SOCIETY.

Non possis oculo quantum contendere Linceus,
Non tamen idcirco contemnas Lippus inungi. Horat. Ep. lib. 1.

LONDON, Printed by *Jo. Martyn*, and *Ja. Allestry*, Printers to the ROYAL SOCIETY, and are to be sold at their Shop at the *Bell* in S. *Paul's* Church-yard. M DC LX V.

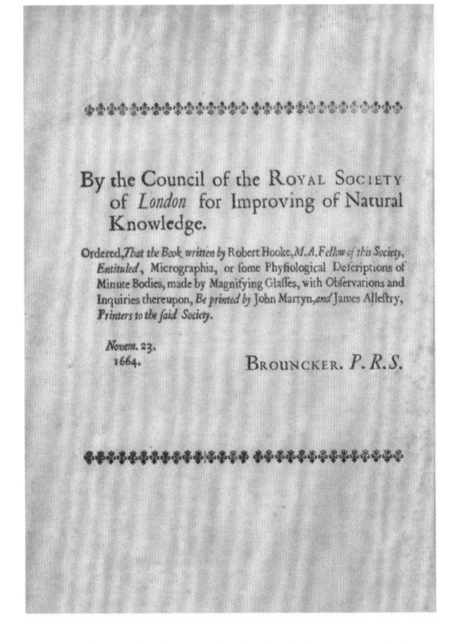

By the Council of the ROYAL SOCIETY of *London* for Improving of Natural Knowledge.

Ordered, *That the Book written by* Robert Hooke, *M.A. Fellow of this Society, Entituled*, Micrographia, or some Physiological Descriptions of Minute Bodies, made by Magnifying Glasses, with Observations and Inquiries thereupon, *Be printed by* John Martyn, *and* James Allestry, *Printers to the said Society.*

Novem. 23.
1664.

BROUNCKER. *P.R.S.*

Pages from Robert Hooke's Micrographia

Ah, the power of a book! When, in 1665, Britain's newly formed Royal Society publishes Robert Hooke's *Micrographia: or, Some Physiological Descriptions of Minute Bodies Made by Magnifying Glasses*, it is an instant bestseller. Hooke (1635–1703), a man of extraordinary talents, has peered through a microscope, drawn pictures of what is visible through it, and put those astonishing drawings in a book.

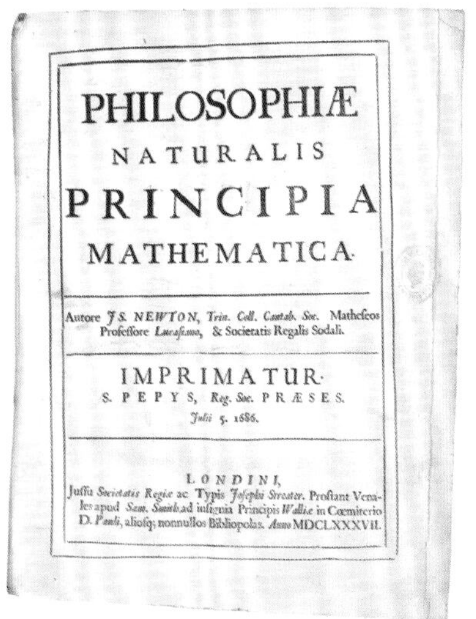
PHILOSOPHIÆ
NATURALIS
PRINCIPIA
MATHEMATICA.

Autore JS. NEWTON, Trin. Coll. Cantab. Soc. Matheseos Professore Lucasiano, & Societatis Regalis Sodali.

IMPRIMATUR.
S. PEPYS, Reg. Soc. PRÆSES.
Julii 5. 1686.

LONDINI,
Jussu Societatis Regiæ ac Typis Josephi Streater. Prostant Venales apud Sam. Smith ad insignia Principis Walliæ in Cœmiterio D. Pauli, aliosq; nonnullos Bibliopolas. Anno MDCLXXXVII.

The title page of Isaac Newton's Principia Mathematica

Britain's Royal Society is founded in 1660 when twelve of London's scientifically minded thinkers attend a lecture by the multitalented Christopher Wren, an architect and professor of astronomy. Those thinkers, who call themselves "natural philosophers," decide to meet regularly to promote what they call "Physico-Mathematical-Experimental Learning." Those meetings turn into the Royal Society, a scientific organization that still flourishes, with headquarters in London and some 1,700 "eminent scientists" as members today.

The society publishes Isaac Newton's *Principia Mathematica*, which becomes foundational in physics, and then a description of how Benjamin Franklin used a kite to show the electrical properties of lightning.

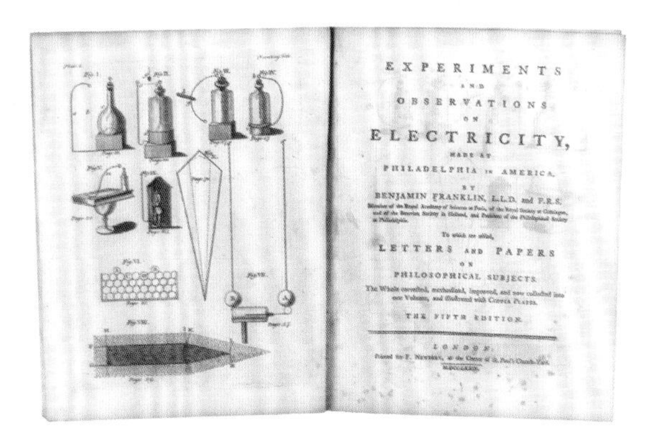
EXPERIMENTS
AND
OBSERVATIONS
ON
ELECTRICITY,
MADE AT
PHILADELPHIA IN AMERICA.
BY
BENJAMIN FRANKLIN, L.L.D. and F.R.S.

To which are added,
LETTERS AND PAPERS
ON
PHILOSOPHICAL SUBJECTS.

THE FIFTH EDITION.

LONDON.

Pages from a 1774 book about Ben Franklin's electricity experiments. These experiments made him famous in Europe's intellectual world.

Regarding Hooke's work, Samuel Pepys (pronounced PEEPS), London's famous diary writer, writes that he saw "Hooke's book of the Microscope, which is so pretty that I presently [bought] it." Pepys starts reading the book and can't put it down. He stays up "till two o'clock in my chamber reading of Mr. Hooke's Microscopicall Observations" and calls it "the most ingenious book that I ever read in my life."

It still dazzles today, in part because of its 117 hand-drawn illustrations, which show a world that can be seen only through a microscope's lens. The book helps spawn an idea that will incubate for a few centuries before leading the way toward much of modern science and philosophy. Here's the idea: the world that we see with our eyes is only part of a much greater and astonishing whole.

Before the seventeenth century, hardly anyone seems to have considered that there is more to the universe than what is visible to the naked eye. And then magnifying glasses that started out as children's toys evolve into powerful tools.

Hooke's friend architect Christopher Wren may have done some of the drawings in *Micrographia*; historians don't agree on this.

Illustrations from Micrographia*: a stinging nettle plant (left) and the eyes and head of a drone fly (right) as seen under a microscope*

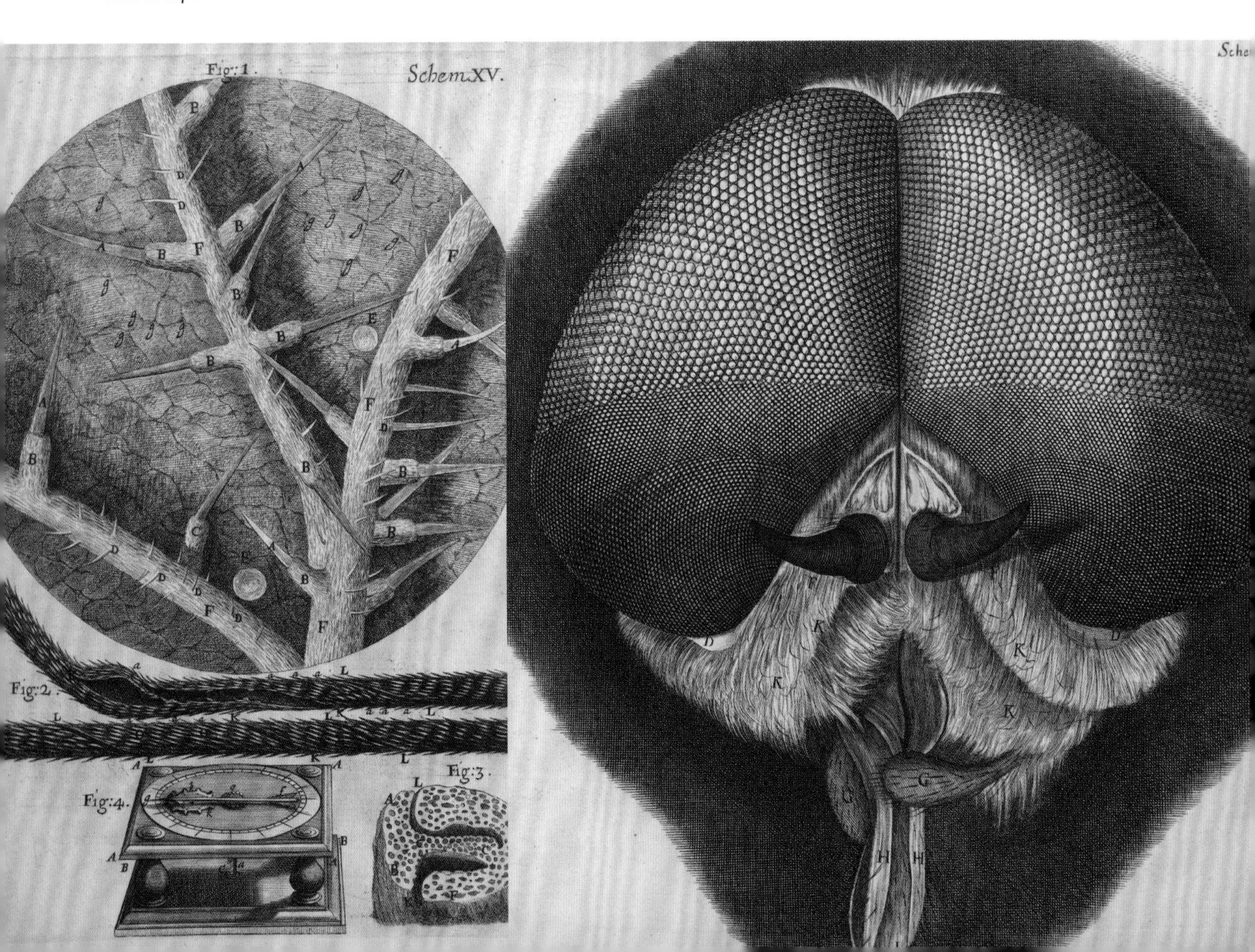

When the curious, like Robert Hooke and some of his peers, begin putting eyes to telescopes, they are amazed to see that there is more in the heavens than the familiar stars and planets. Then, using those telescope lenses in reverse to create microscopes, they discover another unsuspected world. And that's not all. Hooke builds a box with a pinhole that lets in a speck of outside light, which is enough to project an image of whatever is outside the box onto a wall opposite the pinhole. When that image hits the wall, or the back of the box, it is upside down, but a mirror can reverse it. This technological marvel, a camera obscura, or pinhole camera, is much simpler than today's cameras. What it does is let a user control a projected image. Robert Hooke makes the camera obscura into a portable box.

Camera Obscura

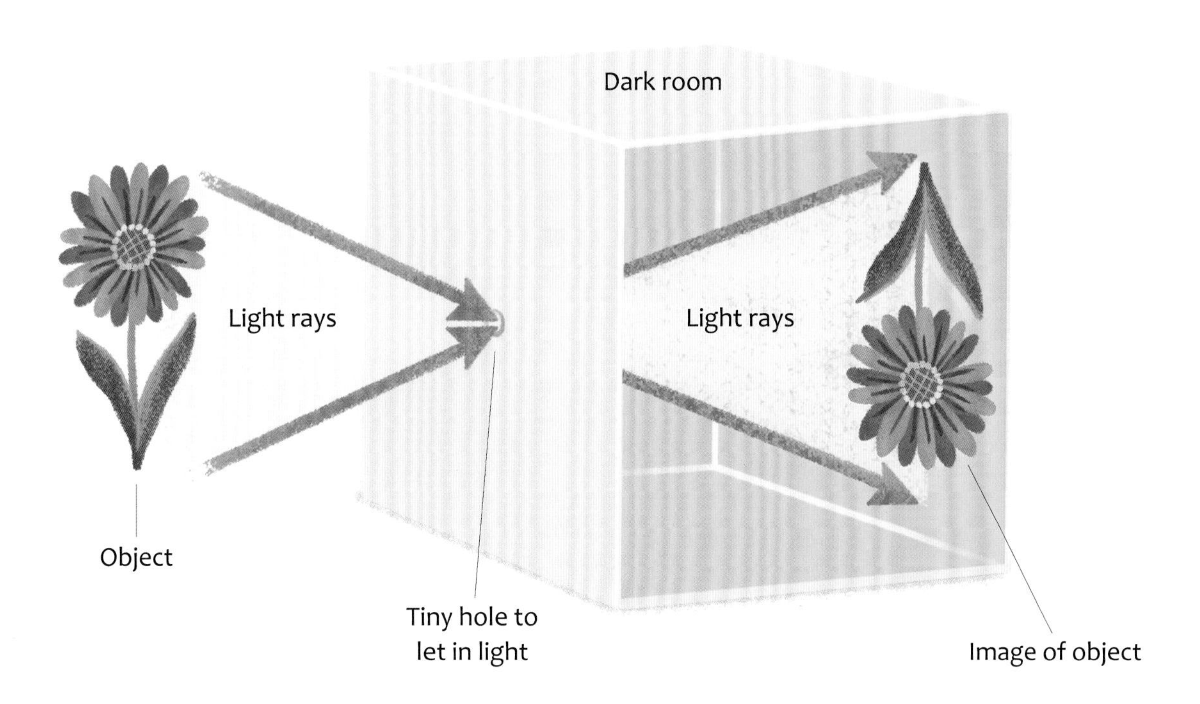

The Chinese philosopher Mozi lived from 470 to 391 BCE. He believed in equality and thought that selfishness is the root of many problems in the world. He also had an interest in science and is the first person known to describe the concept of the pinhole camera. Aristotle, who was born in Greece after Mozi died, also used a pinhole camera, in his case to view an eclipse.

Alhazen (965–1040 CE), an Arab physicist, author, and inventor, made use of pinhole cameras as well. Alhazen lived in Islam's Golden Age, which is usually dated from the eighth to the thirteenth centuries. Building on the work of Aristotle, Alhazen helped pioneer what we now call the scientific method, based on confirmable experiments and mathematical evidence. His *Book of Optics* was translated into Latin and studied in Europe's universities. Alhazen understood that if you want to watch a solar eclipse without damaging your eyes, you can do that if you observe it as a reflection on a wall that has come through a tiny pinhole. When that hole is in a box, the box becomes a camera obscura.

An illustration showing some properties of light from a 1572 Latin edition of Alhazen's Book of Optics

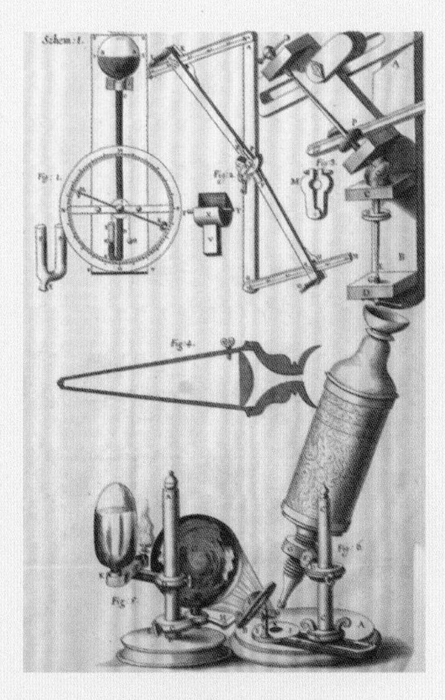

An engraving of Hooke's microscope from his book Micrographia

Christoph Scheiner (1573–1650), a German Jesuit priest fascinated with technology, physics, and astronomy, invented a pantograph (a device for copying drawings) and a helioscope (pictured here), a device for observing the sun.

The title page illustration from a 1733 book, Osteographia*. It shows a camera obscura being used to project an image of a partial human skeleton; inside the camera, an artist is using the image as the basis for a drawing. This method was used to create the book's many strikingly detailed drawings.*

Who actually invented the pinhole camera? No one knows. It seems to have been in use going back to ancient times.

In seventeenth-century Europe, lenses have become more than a curiosity. For anyone with imagination, they are the future. In Holland, a scientist named Christiaan Huygens (1629–1695) develops a variation on Hooke's box. He makes it bigger, has it project images from slides onto a wall, and calls it a "magic lantern." These magic lanterns are inexpensive technological marvels that can be made by any savvy teenager. Huygens's magic lanterns will eventually lead to motion picture projectors.

We in the twenty-first century can appreciate the lure of technology, and so did people, especially young people, in the seventeenth century. The magic lantern attracts innovative thinkers. And that brings us back to Britain's Royal Society, where in 1660, twelve gentlemen who are scientific explorers—or, in the terminology of the time, "natural philosophers" (philosophers of nature)—make plans to meet weekly. Among them is twenty-eight-year-old Hooke, who is asked to serve as "curator of experiments" for the newly formed Royal Society. They take Francis Bacon's well-known words "Knowledge itself is a power" as a kind of credo.

Hooke is a polymath—a mathematician, architect, organist, astronomer, scientific theorist, and surveyor. Besides all those thinking skills, Hooke is also able to make things with his hands, including a clock of wood. One of the society's other members, John Aubrey, says of him, "He is of prodigious inventive head . . . a person of great vertue and goodness. . . . He is certainly the greatest Mechanick this day in the World."

But Hooke, from a working-class background, is vulnerable and thin-skinned. He will get into a nasty fight with the society's future president, Isaac Newton, who has one of science's greatest minds but is touchy and unforgiving.

Hooke's microscope

LEFT: This 1737 etching by Anne Claude Philippe de Tubières, comte de Caylus, shows a street musician carrying a hand-cranked organ strapped to her front and a magic lantern, a device based on the camera obscura, strapped to her back.

Hooke, in his lectures, introduces ideas Newton will then develop. Maybe that helps explain why these two amazing scientists never get along. Still, it is the five-year-old Royal Society that publishes Hooke's great book *Micrographia* and Isaac Newton's monumental *Principia*.

Hooke has spent much time peering through lenses. His wonderful book contains carefully drawn pictures of a fly's eye, a louse, a flea, pieces of linen and silk, the point of a needle, and much more. Hooke does something else that's new: he takes a very thin section of cork and, with a mirror, shines light through the specimen, thereby using transmitted rather than reflected light, which will pave the way for some of the microscopes to come.

Half a century after microscopes come into use as scientific instruments, people are using them to discover an "infinity of living animals" in water, vinegar, milk, cheese, blood, and many other substances.

In 1661, Carlo Antonio Manzini writes in *The Conquests of the Microscope*:

> *And now thanks to microscopes (these are little telescopes, which enormously enlarge small objects put under the lens) . . . it will be possible to conduct more logical discourse on the marvels of nature. They agree with me, those who have by this means discovered in very strong vinegar, living and darting about, innumerable tiny eels or snakes; and in milk, in the blood of feverish patients, and in powdered cheese, an infinity of living animals; and almost in competition with the aforesaid 'occhiale' of Galileo, discoverer of innumerable little stars mentioned by Democritus but not seen, in the Milky Way in Heaven, have discovered on Earth just as many living souls that were unknown to men, whose eyes were not made by nature keen enough to discover those tiny bodies.*

RIGHT: An illustration from Robert Hooke's Micrographia *showing slices of cork viewed through a microscope*

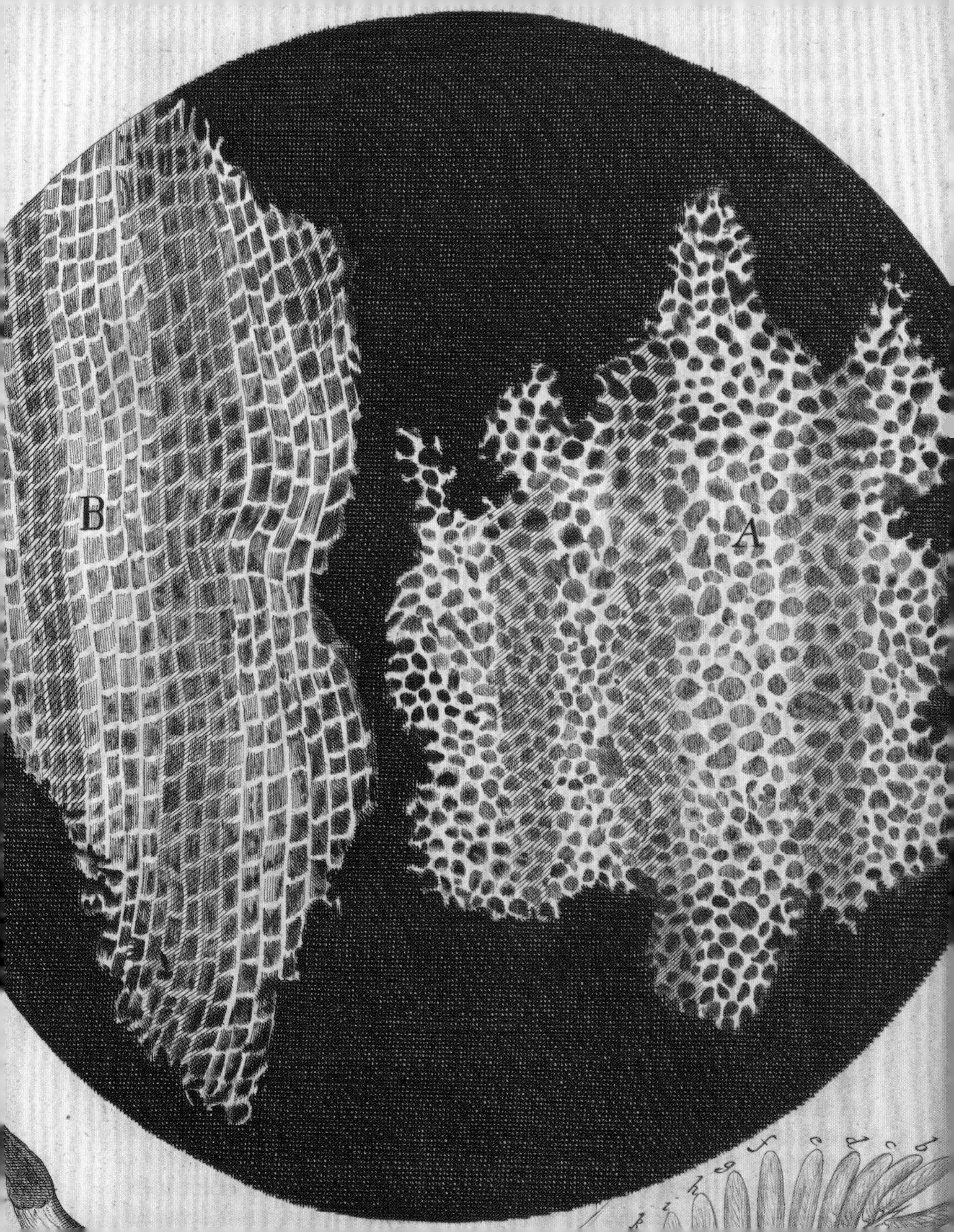
B
A
f
e
d
c
b
g
h
i
k

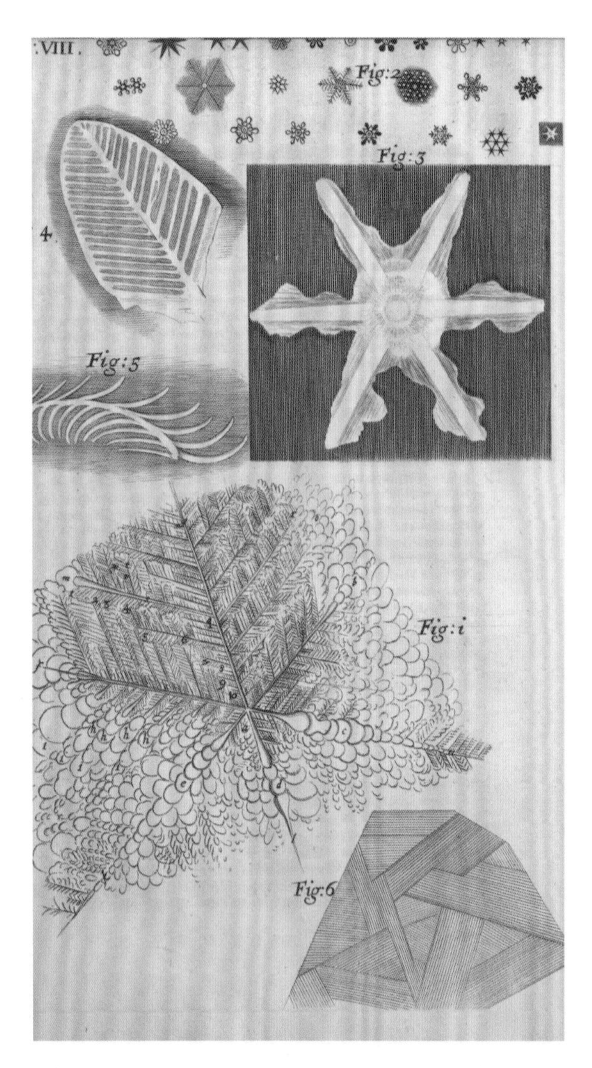

A page from Hooke's Micrographia *showing snowflakes*

Examining a piece of cork, Hooke can see that it is divided into tiny chambers that remind him of monks' chambers, so he calls them cells, creating a scientific term.

He comes up with the figure 1,259,712,000 as the number of cells in a one-inch square of cork. For the first time in recorded biology, a gigantic number is used appropriately. Not only does Hooke calculate that the square of cork contains more than a billion cells; he also provides the foundation for what will become cell theory, or the idea that cells are the basic unit of life.

As for the flea, the gnat, and the mite under his lens, Hooke and his Royal Society colleagues are astonished to find that they share basic body parts with horses, elephants, and lions. This is a breathtaking insight. It isn't always easy, even for Hooke, to use the microscope. He writes that an ant was "troublesome to be drawn," so he gave it some brandy (don't try this at home, please), which "knock'd him down dead drunk, so that he became moveless." After an hour, the ant "suddenly reviv'd and ran away" with "several bubbles" coming out of its mouth.

Using a compound microscope takes real skill. Samuel Pepys buys his from a Mr. Reeve at "a great price, but a most curious bauble it is . . . and he makes the best in the world." Then Pepys and his wife try to see something through the lens. They do so "with great difficulty." In fact, they can't see much of anything and go back to admiring Hooke's *Micrographia*.

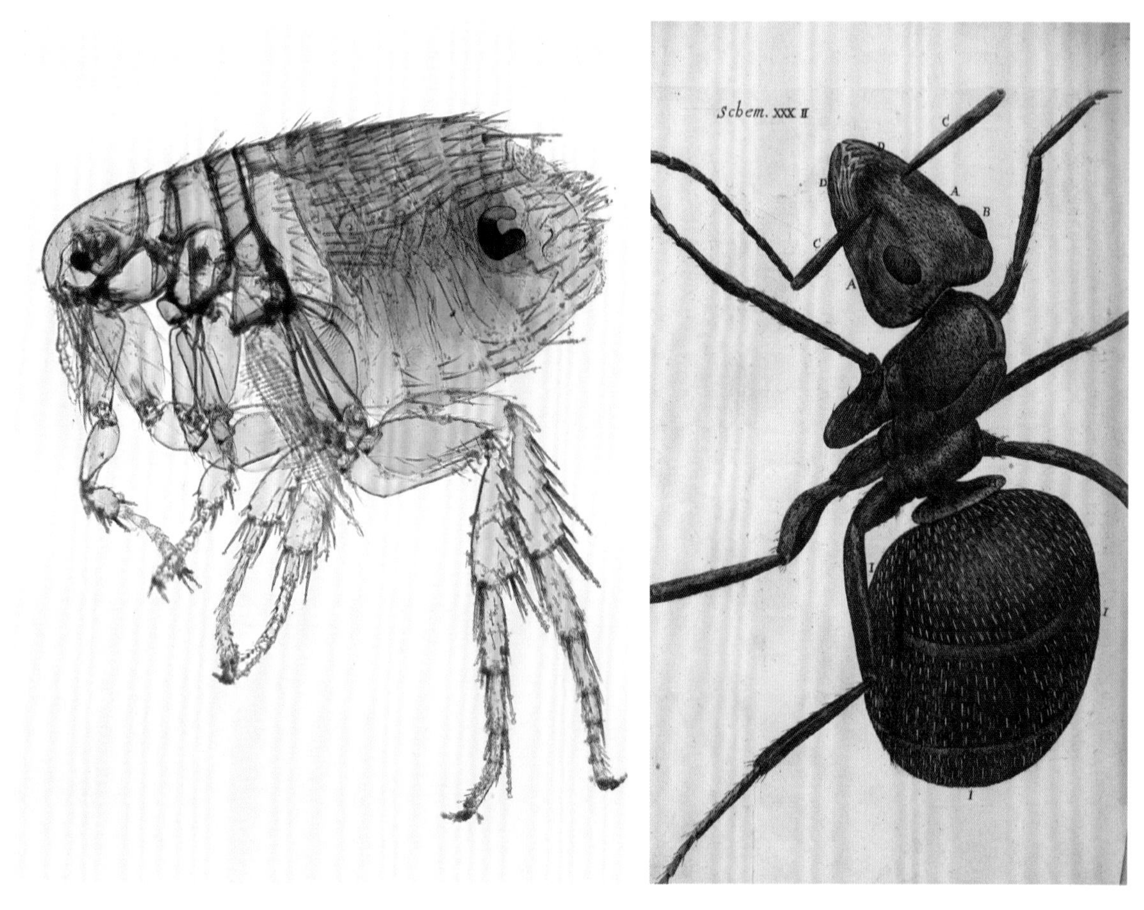

A modern magnified image of a flea (left) and an illustration from Hooke's Micrographia *of an ant as seen through a microscope (right)*

Neither Mr. and Mrs. Pepys nor Hooke realize that the small creatures they are considering are huge when compared with still smaller entities that, centuries later, will be known as bacteria and archaea and viruses. Hooke's microscope doesn't have the power to see into the world of those minute life-forms. No one in Hooke's time knows that Earth holds a world of microbes—tiny one-celled creatures with numbers so vast that their combined biomass is larger than that of all the living creatures visible to our eyes. ■

Arcana
NATURÆ
detecta
ab
ANTONIO
van
LEEUWEN
HOEK.
J. van Schaak fecit.
occult

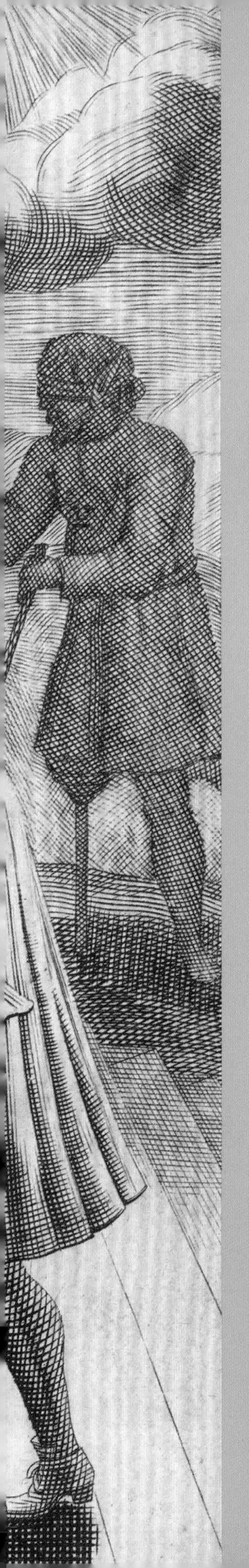

08

Seeing More Is Better: Enhancing Magnification; Using an Artist's Eyes

Our universe is a sorry little affair unless it has in it something for every age to investigate. . . . Nature does not reveal her mysteries once and for all.

—Seneca

We may hope that amateurs in the coming century, using the new tools that modern technology is placing in their hands, will invade and rejuvenate all of science.

—Freeman Dyson

LEFT: Part of the title page of the 1695 book Arcana Naturae Detecta *by Antonie van Leeuwenhoek, a Dutchman who became famous for what he saw through his lenses. Note the woman in the center looking through a magnifying lens.*

It is 1668, eight years after the Royal Society's founding, and a thirty-six-year-old merchant is making his first and only visit to London. Antonie van Leeuwenhoek (1632–1723) (say "LAY-ven-hook" and you'll be close) doesn't speak much English and can't read Latin, the language of the scholarly world. He has come from Delft, a small Dutch city known for beer, tapestry, and pottery. While he is in London, he makes sure that he sees Hooke's *Micrographia*, which is the talk of the city. It includes magnified pictures of textiles (his special interest) and of insects that are astonishing. The book will change his life's direction.

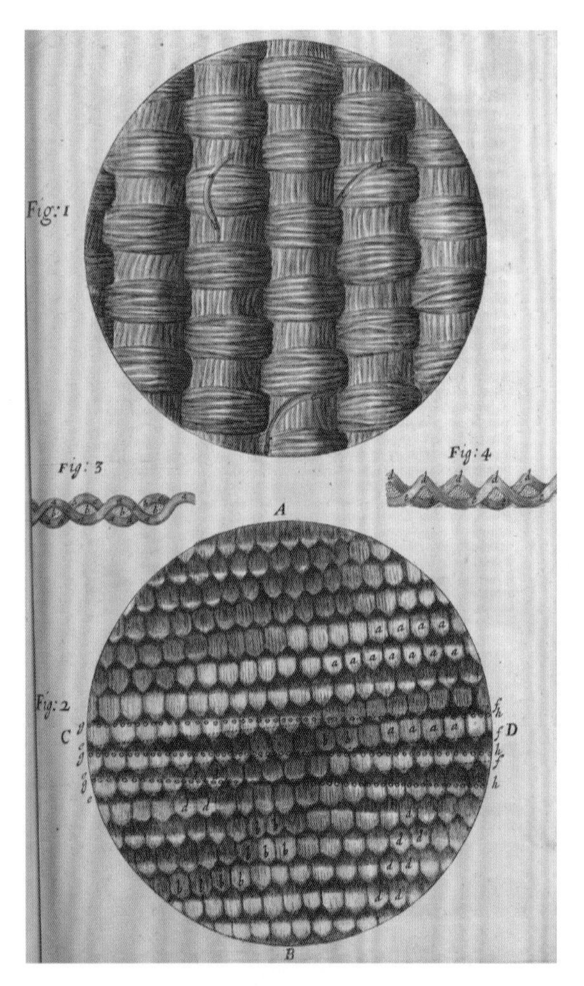

Illustrations of silk viewed through a microscope, from Hooke's Micrographia

Leeuwenhoek is a certified master in cloth-making. As a fabric expert, he is expected to count threads in cloth. To do that, he uses a magnifying glass, which at the time is a high-tech device.

Leeuwenhoek begins his career as an apprentice to a fabric master in Amsterdam, which is where he learns to use magnifying devices. When Leeuwenhoek returns to his hometown from Amsterdam, he opens a fabric shop. He and his family live above the shop, which prospers. He inherits money from his mother's family and becomes a man of substance: a surveyor, chamberlain to the sheriffs of the city of Delft, and an official wine taster.

So by the time he goes to London, he is financially independent and able to do what he wants with his life. And what Leeuwenhoek wants to do is peer through a lens and see what he can discover.

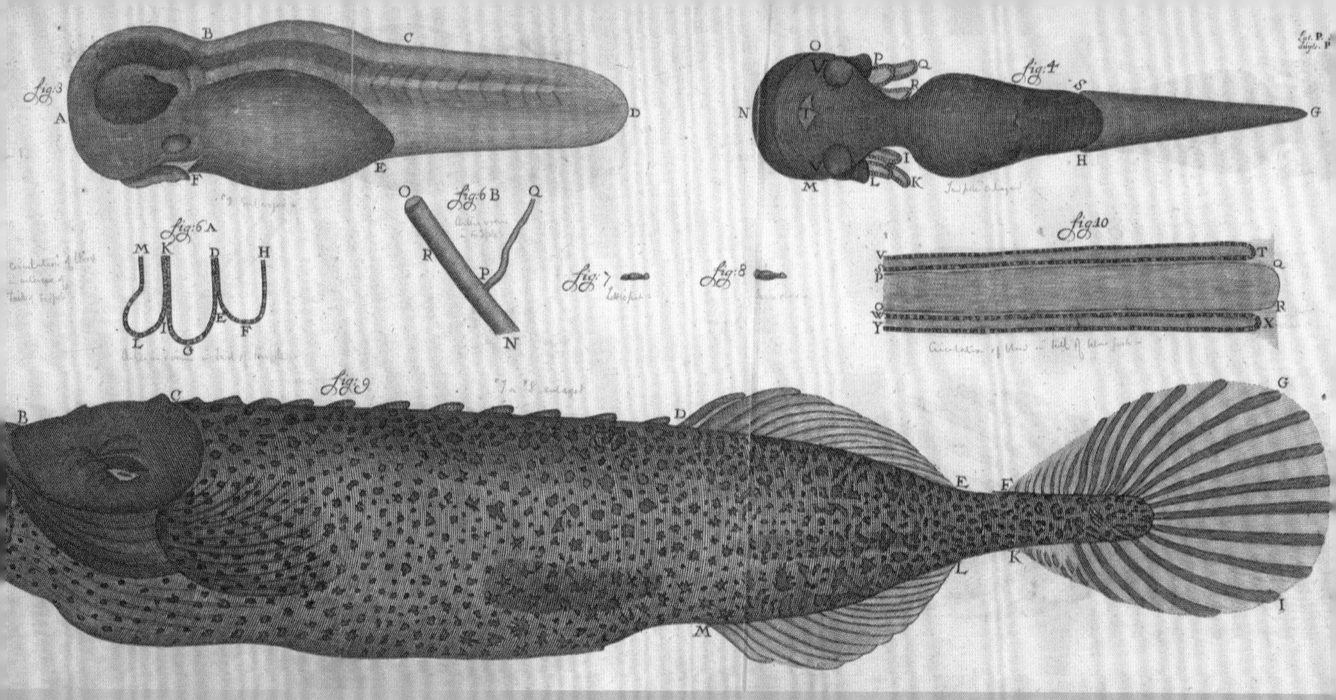

Illustrations of frog and fish specimens as viewed under a microscope from Leeuwenhoek's Arcana Naturae Detecta

As Leeuwenhoek becomes serious about pursuing microscopy beyond the fabric world, he seeks out experts in other fields where his microscopes might prove revelatory. Jan Swammerdam is one of them.

Trained as a physician, Swammerdam (1637–1680) has access to corpses and becomes well known for his own research with microscopes; he does dissections and makes discoveries in anatomy, specifically insights into respiration and nerve-muscle function. Swammerdam focuses his lenses on insects and finds that eggs, larvae, and pupae are not different kinds of life but all forms of the same creature in various stages of development. He becomes aware that insects incubate in an egg laid by a female of its species. That doesn't fit the beliefs of most seventeenth-century religious thinkers, who say that insects arise at God's command.

Inspired by the beauty of the life he can see under the microscope yet unable to handle its religious implications, Swammerdam finds himself pulled between his scientific and religious impulses and struggles with depression. But his *Bible of Nature*, a collection of microscopic observations, will become a classic in the field.

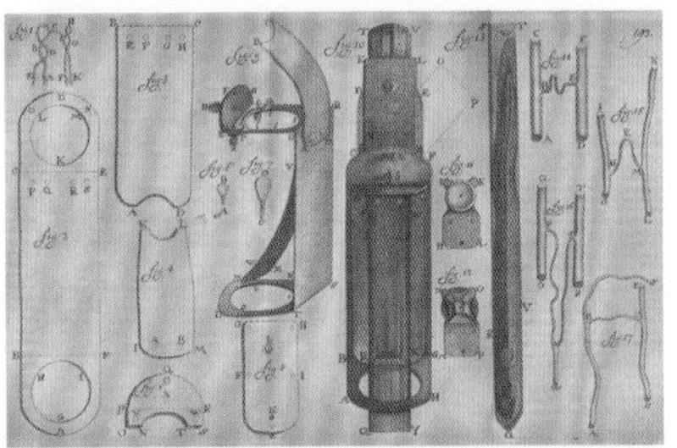

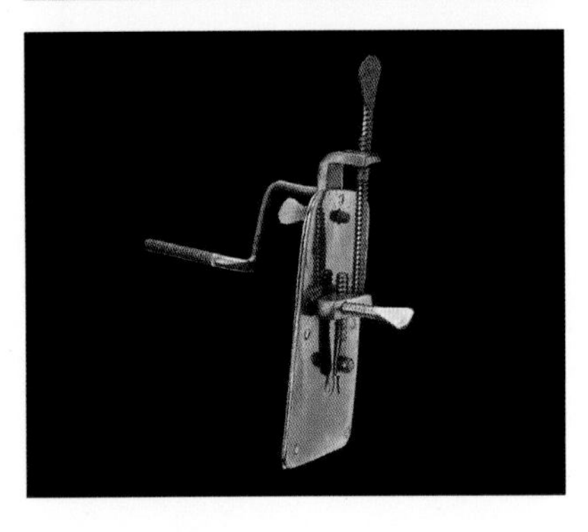

A 1669 painting by Johannes Vermeer called The Geographer *(top); a diagram of a microscope from Leeuwenhoek's* Arcana Naturae Detecta *(middle); a replica of a Leeuwenhoek microscope from the Smithsonian (bottom)*

In this time, the Dutch Republic provides the intellectual freedom that makes exploration possible, whether it is in trade, the arts, or the world of microbes. Holland is ruled by state governors instead of a king. It is a tolerant society, where women and religious outcasts from other countries, like Britain's Pilgrims, are shown respect. Freedom pays off. The middle class prospers. The Dutch East India Company becomes the largest shipping company in the world, and the University of Leiden is the premier institution for science in Europe. Artists like Rubens, Rembrandt, and Vermeer are free to paint masterpieces. They are among those laying foundation stones for a new intellectual edifice supported by a burgeoning publishing world and an abundance of books, newspapers, and literary salons. A rise in literacy will enable ordinary people to challenge new ideas and shape the course of philosophy, art, and science.

Leeuwenhoek is a genius, and he does things his own way. Hooke's microscopes, at their best, enlarge objects twenty or thirty times. Leeuwenhoek builds a simpler but more powerful device. He starts with a single lens held in a brass plate. A specimen is kept in place by sticking it on a sharp point, where its position can be shifted by the viewer. The device is held close to the eye. It demands an exquisite lens and a skilled viewer.

Leeuwenhoek is myopic, or nearsighted, which means his eyesight is very strong when he looks at objects at close range. This gives him an advantage in microscopy. Soon he is magnifying objects more than two hundred times and seeing

Vermeer and Leeuwenhoek

A portrait of Antonie van Leeuwenhoek from Arcana Naturae Detecta

Does Leeuwenhoek hire Vermeer to make drawings? It's possible. The artist Johannes Vermeer (1632–1675), who is the same age as Leeuwenhoek, lives nearby in Delft and has eleven kids to feed.

Vermeer is doing paintings that will rank among the world's greatest, and in 1662, he will be named head of the Delft painters guild. No one is sure, but Vermeer might even be using a camera obscura, the device Robert Hooke has popularized, to give detail to his paintings. Vermeer's family, like Leeuwenhoek's, is part of the fabric trade, where magnifying skills are expected.

Could the young artist and the young microscopist share ideas about trendy technology? Both are outsiders in the educated world of their time. As far as we know, Vermeer is not part of the studio of a renowned master, and Leeuwenhoek, who can't read Latin, is not part of the schooled elite. Yet both will become world-class figures in their chosen fields. Leeuwenhoek will use a magnifying device as no one seems to have done before. Vermeer will paint with detail and precision and an understanding of light that few others achieve.

We don't know how well the two men knew each other, but we do know that when Vermeer dies, Leeuwenhoek is asked to handle his family's affairs. Fame will come to Vermeer after he is gone; Leeuwenhoek will be a figure of awe in his time.

things never seen before. He lets friends take a peek through his lenses, and he also hires an illustrator to draw what can be seen through them.

A prominent doctor in Delft, Reinier de Graaf, sends a letter to Henry Oldenburg, the secretary of London's Royal Society. He says:

> *I am writing to tell you that a certain most ingenious person here, named Leeuwenhoek, has devised microscopes which far surpass those which we have hitherto seen. . . . The enclosed letter from him, wherein he describes certain things which he has observed more accurately than previous authors, will afford you a sample of his work.*

Leeuwenhoek's letter carries drawings that depict a fungus, the eye and mouthparts of a bee, and the whole body of a louse picked out of someone's hair. Could these extraordinary pictures be works of imagination? The Royal Society wants to know. Constantijn Huygens (1629–1695), a Dutch mathematician and astronomer who has been knighted by King James I of England, writes a character reference:

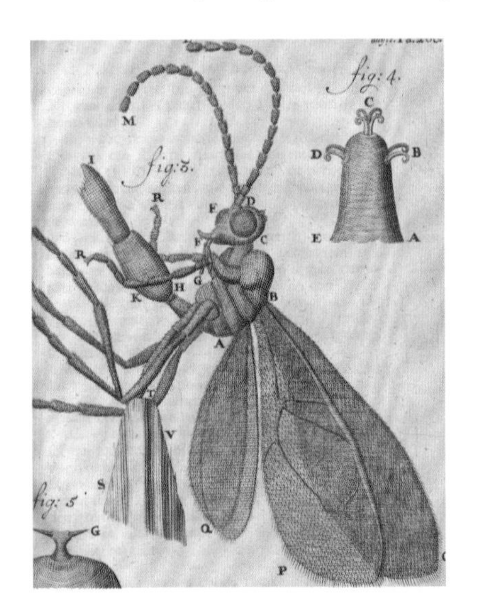

A drawing of a wasp as seen under a microscope, from a letter by Leeuwenhoek to the Royal Society in London

> *Our honest citizen, Mr. Leeuwenhoek . . . unlearned both in sciences and languages, but of his own nature exceedingly curious and industrious. . . . I trust you will not be unpleased with the confirmations of so diligent a searcher as this man is.*

The Royal Society publishes the letters and pictures in its journal. Oldenburg, who can read Dutch, translates Leeuwenhoek's descriptive commentary.

Leeuwenhoek has been focusing on butterflies, fleas, and other insects studied by anatomists of the day. Then he finds his own path. Using his more powerful lens, he examines fungi on cheese, bile from animals, blood from a finger, crystals formed in urine, and water of all kinds: ditch water, river water, seawater, rainwater, and snow.

He writes with clarity: "The globules in the blood [are] as sharp and clean as . . . sand grains that one might bestrew upon a piece of black taffety silk." He has discovered red blood cells.

When he looks at sterile water, he finds nothing in it. Then he leaves the water in an open container in his room. He writes, "I had no thought of finding any living creatures in it; but upon examining it, I saw with wonder quite 1,000 living creatures in one drop of water."

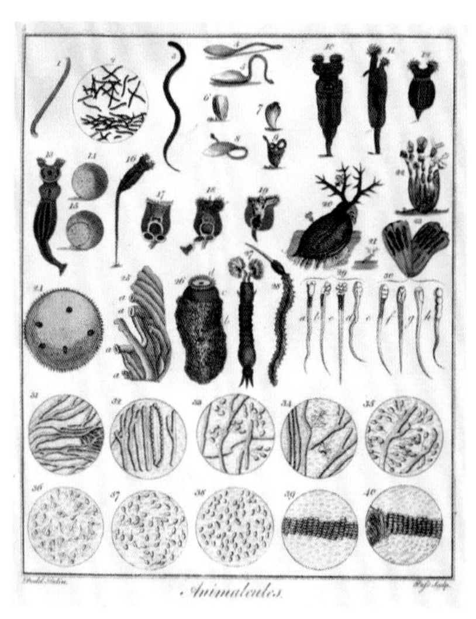

"Animalcules," or microscopic animals, observed by Leeuwenhoek

Twenty-four hours later he writes, "Observing again, I saw the foresaid animalcules in such great numbers in the water . . . that now they did not amount to merely one or two thousand in one drop."

The next day, he writes that some of them "swam gently among one another, and moved after the fashion of gnats in the air," while larger ones "had a much swifter motion; and as they turned and tumbled all around and about, they would make a quick dart."

Some of the creatures, he writes, were "so small, in my sight, that I judged that even if 100 of these very wee animals lay stretched out one against another, they could not reach to the length of a grain of coarse sand."

It is October 9, 1676, and the animalcules that Leeuwenhoek describes are bacteria. As far as we know, he is the first person to see them and other tiny life-forms.

Bacteria are single-celled organisms that come in a variety of shapes. Bacterial cells have no nucleus, which is a core part of many other kinds of cells.

When he scrapes plaque from his teeth, Leeuwenhoek describes what he sees as "a little white matter, which is as thick as if 'twere batter." Seen through a lens, that white matter looks to be alive. Leeuwenhoek finds two men in Delft whose teeth have never been brushed. In the mouth of one

e
f
i
k
b
g
a
h

Maria Sibylla Merian (1647–1717) is born in Frankfurt, Germany, into a family of publishers, printers, and artists. Her father, a well-known Swiss publisher, dies before she can know him; her stepfather, Jacob Marrel, is an artist who teaches painting to classes of boys. Growing up, Maria is able to join the boys. Her stepfather's specialty is painting flowers, and sometimes he adds insects to those paintings. He encourages his students to do the same thing. Merian is soon painting flowers with a vibrancy and skill that go beyond those of most artists of her time. Then she goes further, as the flies and beetles she includes in her paintings capture her attention; she begins to carefully study them as a scientist would.

Her fascination with bugs becomes a passion. She turns the study of the metamorphosis of silk-worms (from egg to larva to pupa to butterfly) into a life focus. In 1679, she publishes *The Caterpillars' Marvelous Transformation and Strange Floral Food*, the first of two volumes in which Merian describes and illustrates (with fifty engraved plates) the caterpillar-to-butterfly transition.

It is both a gorgeous book with superb examples of art and a scholarly book of explanatory science.

For a while, Merian leads a conventional life. She is married and the mother of two daughters, both of whom paint in her style. But her marriage fails. She divorces, moves from Frankfurt to Amsterdam (Holland's vibrant capital city), and joins a newly formed Christian community, the Labadists, who share possessions and believe in the equality of men and women.

LEFT: An illustration by Maria Sibylla Merian depicting a variety of flowers as well as a beetle and two flying insects. ABOVE: A watercolor by Maria Sibylla Merian showing the life cycle of the vine sphinx moth

Then she does something that takes courage in an era when it is unusual for women to travel together independently. In 1699, at age fifty-two, she and her twenty-one-year-old daughter, Dorothea, sail across the Atlantic Ocean to Suriname, a lush tropical land on the east coast of South America, where a colony of Labadists has settled.

Merian is fascinated by descriptions of the plant and animal life that she has heard about from Labadists who have been to Suriname and returned to Europe.

"In Holland I marveled to see what beautiful creatures were brought in from the East and West Indies," she will write later. "This prompted me to under-take a long and expensive journey."

To sail to Suriname takes two months in a small boat. When Merian arrives, she is in a scientific wonderland. Focusing her artist's eyes on plants and bugs and other small animals, she begins creating a body of paintings that will be turned into a book. Titled *Metamorphosis Insectorum Surinamensium*, it will be published in 1705.

Spiders and ants from Maria Sibylla Merian's 1705 book, Metamorphosis Insectorum Surinamensium

As she describes it:

> *The work consists of sixty copperplate engravings on which are displayed some ninety studies of caterpillars, worms, and maggots; how they change in color and form when molting, and finally change into butterflies, moths, beetles, bees, and flies. All these creatures are shown on the same plants, flowers, and fruits they ate for their nourishment. Here are also included life stages of West-Indian spiders, ants, snakes, rare toads and frogs.*

A watercolor by Maria Sibylla Merian that shows a caiman holding a false coral snake in its mouth

of them is "an unbelievably great company of living animalcules, a-swimming more nimbly than any I had ever seen up to this time. The biggest sort . . . bent their body into curves in going forwards." All the spit, he added, "seemed to be alive."

He puts semen under the microscope and watches sperm swim; he believes this is one of his most important discoveries. And he is right. When he looks at spermatozoa (sperm cells) from mollusks, fish, birds, and other animals, he decides that fertilization occurs when the sperm penetrates the egg (he is right again). He goes further and says that the sperm is the embryo-to-be, the female egg being just a carrier in the reproduction process (this time he is wrong).

Spices have helped make the Dutch rich. But what makes pepper hot to the taste? Leeuwenhoek guesses that pepper grains must have sharp thistle-like projections that sting the tongue. He can't find any. What he sees when he soaks hot peppers in water is astounding: thousands of single-celled creatures are moving vigorously.

What is it that makes peppers hot? Peppers contain the chemical capsaicin, which binds to receptors that are in nerve endings on your tongue and in your throat. The chemical makes tissue it comes into contact with feel like it is burning. And, yow, that can hurt.

Leeuwenhoek becomes famous. Russia's Peter the Great is among those who come to Delft just to see him. His town, proud of its son, gives him a pension so he can focus on his lenses.

Of course he is flattered by the attention, but in a letter to a friend he complains about the interruptions and says he would prefer to be left in peace to do his work. His only intent, he says, is "to go over to the Truth, and to cleave unto it." In a letter to the Royal Society in 1716, he writes:

> *My work, which I've done for many a long year, was not pursued in order to gain the praise I now enjoy, but chiefly from a craving after knowledge, which I notice resides in me more than in most other men. And therewithal, whenever I found out anything remarkable, I have thought it my duty to put down my discovery on paper, so that all ingenious people might be informed thereof.*

He sees it as his duty to share his discoveries. There's grandeur in that thought. Is scientific achievement a product to be sold or a resource that belongs to everyone? For Leeuwenhoek the answer is clear.

For fifty years, he sends drawings and observations to London. Oldenburg translates the words; the Royal Society publishes most of them. At ninety, dying of a rare malady (now called Leeuwenhoek's disease), Leeuwenhoek studies and describes his illness in exact detail.

For years to come, historians will describe him as a lens grinder with no peer. Only recently have we learned that lens grinding wasn't the secret of his success, rather it was glassmaking. In a recent study, Dutch scientists used a neutron beam to take three-dimensional computer images of some of Leeuwenhoek's surviving microscopes and confirmed an unusual glassblowing method he used to make his unique microscopes. It was the design of his apparatus and the shape of his lenses, rather than how he ground the glass in his lenses, that made a difference. Leeuwenhoek took a rod of soda lime glass, heated it over a flame, and pulled it apart, creating threads of glass. When he put the end of a thread back in the flame, he got a very small glass sphere. From these spheres, he created lenses for his microscopes. The greatest magnification came from the smallest spheres. ■

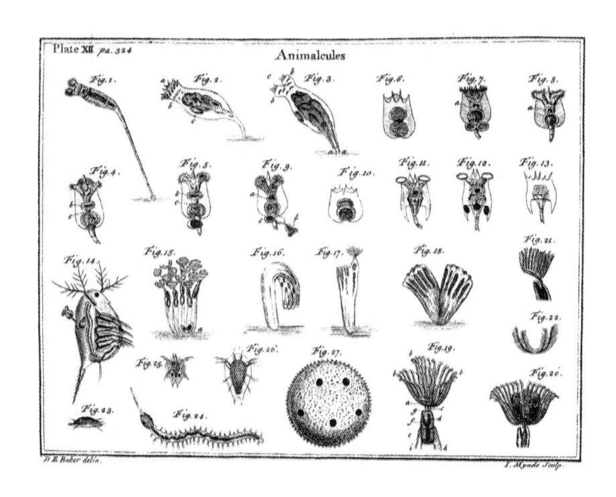

A 1753 depiction of animalcules by Henry Baker, an English naturalist

Have you ever played the game Twenty Questions and started out by asking, "Animal, vegetable, or mineral?" Well, protists are none of those. The cells of a protist have a nucleus, but a protist is not a plant, animal, or fungus. What is it? Actually, there are differing opinions about that. Usually, a protist is defined as a single-celled organism so small that you need a microscope to see it.

P. J. Redouté pinx. Imprimerie de Rémond Langlois sculp.

09

The Name Game: Linnaeus Tries to Name Everything

What's in a name? That which we call a rose
By any other name would smell as sweet.

—William Shakespeare

I pull a flower from the woods—
A monster with a glass
Computes the stamens in a breath—
And has her in a "class"!

—Emily Dickinson

LEFT: A centifolia rose, painted by Pierre-Joseph Redouté

Most seventeenth-century Swedes don't have a family name; they use their father's name to identify themselves. Which means that Nils Ingemarsson is the son of an Ingemar. His father, Ingemar Bengtsson, is the son of a Bengt. And Bengtsdotter is a daughter of a Bengt. This can lead to confusion: there are a lot of Ingemarssons and Bengtsdotters in Sweden.

So when Nils Ingemarsson enrolls in Uppsala University, he is asked to pick a new surname. He chooses to name himself after a linden tree on his family property, Linnågard, and then Latinizes it to Linnaeus. Why Latin? Having a Latin name is a sign that you are educated. For scholars throughout the Western world, Latin is a shared language.

Before there was DNA testing, if you wanted to know something about your family history, you usually focused on surnames. Tracking family names can be surprising and sometimes fun. Here's an example: The name Clark comes from the Latin *clericus*, which means "cleric," or "scribe." But in Old English it was *clerec* and usually meant "priest." So if you know a Ms. Clark, you can guess that someone in her family may have been a member of the clergy or a scribe.

Nils, who becomes a country pastor, speaks Latin at home to his children, so his son Carl (1707–1778) is raised in a bilingual home. It's also a home where he is surrounded by plants and animals and their lore. That's not unusual in the eighteenth century, a time when the educated public is captivated by nature. One reason: if you discover a new species, you get to name it. So backyard explorers, as well as academics, do a lot of sleuthing in woods and waterways and other natural environments. Some plants get a bunch of names, which, of course, is confusing.

Carl Linnaeus attempts to make sense of the naming process, and in doing so he becomes a celebrity. (His primary interest is in plants, since he is a botanist, but he will also name animals.) A fellow Swede, author August Strindberg, says that Linnaeus "was in reality a poet who happened to become a naturalist."

Linnaeus writes:

> *For riches vanish, the most stately mansions fall into decay, the most prolific families die out sooner or later: the mightiest states and the most flourishing kingdoms may be overthrown: but the whole of nature must be obliterated before the genera of plants disappear and he be forgotten who held the torch aloft in botany.*

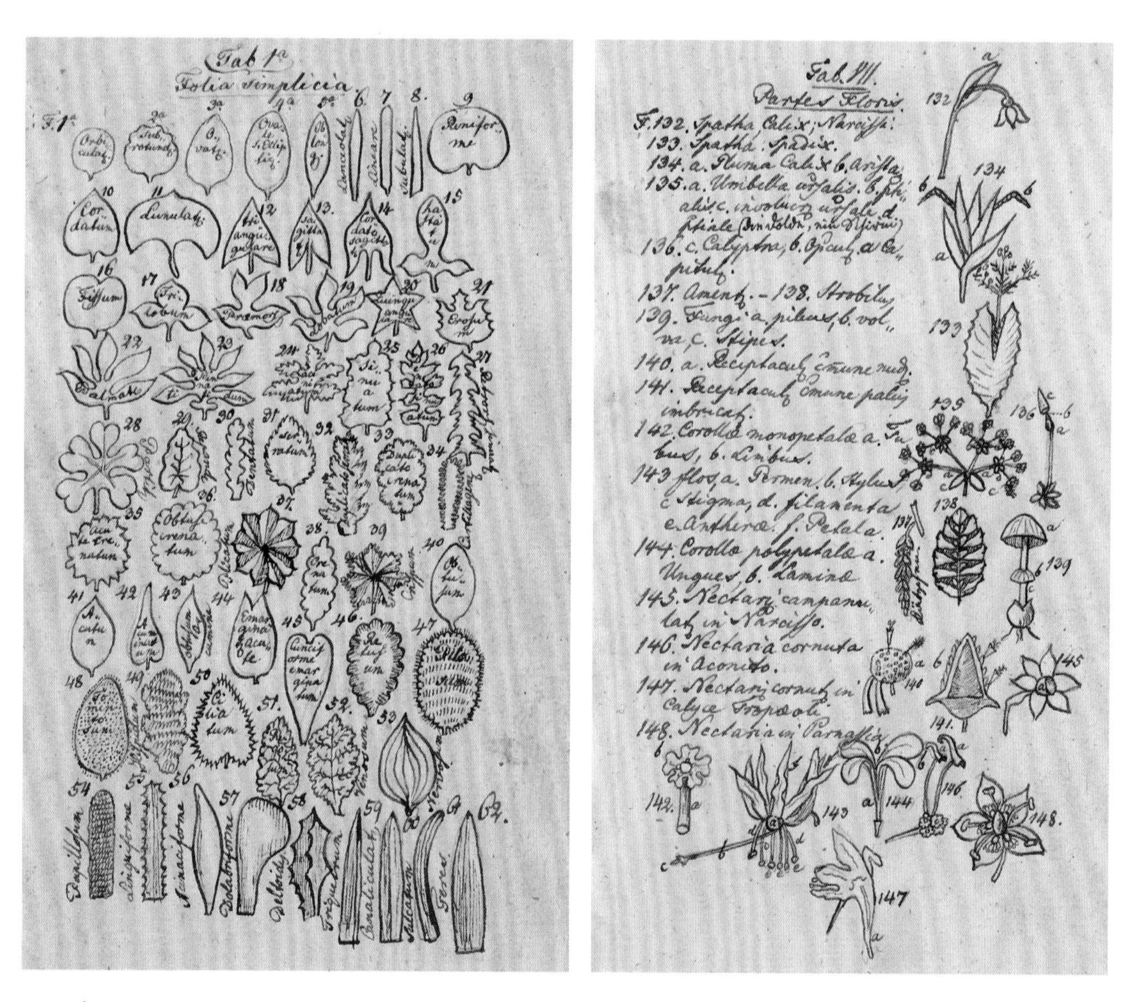

Pages from Linnaeus's notebooks that include sketches of leaves and parts of flowers

Meanwhile, in the Americas, an Outbreak of Smallpox

Onesimus, an enslaved Black man, lives in Boston, Massachusetts, at a time, in the eighteenth century, when slavery is tolerated there. Onesimus will make an impact on the world of medicine and public health that sends waves beyond the Massachusetts colony.

Onesimus was purchased for Cotton Mather, an important Puritan minister, by members of Mather's congregation—an indication of how widely accepted slavery was at the time. Mather is best known today for his role in a famous trial of people accused of being witches.

This is an era when smallpox, a contagious disease, is a scourge that often kills. When it doesn't, it usually leaves scars behind. Most doctors in the eighteenth century don't know how

A 1728 engraved portrait of Cotton Mather

TO BE SOLD, on board the Ship *Bance-Yland*, on tueſday the 6th of *May* next, at *Aſhley-Ferry*; a choice cargo of about 250 fine healthy

NEGROES,

juſt arrived from the Windward & Rice Coaſt. —The utmoſt care has already been taken, and ſhall be continued, to keep them free from the leaſt danger of being infected with the SMALL-POX, no boat having been on board, and all other communication with people from *Charles-Town* prevented.

Auſtin, Laurens, & Appleby.

N. B. Full one Half of the above Negroes have had the SMALL-POX in their own Country.

A newspaper ad from 1760 announcing an auction near Charleston, South Carolina, of Africans taken in bondage. A note indicates that half have already had smallpox.

to treat it. If you catch the virus, police may nail the door of your house shut; sometimes authorities put a red flag on a home with infected people to warn others to stay away. Sometimes they add a sign that says *God Have Mercy on This House*.

Onesimus knows things about smallpox that no one else in Boston seems to know. He tells Mather that he was inoculated against smallpox when he was a child in Africa. A bit of pus from someone with smallpox was rubbed on a cut in his skin, which gave him a mild case of the disease and lifetime immunity.

Cotton Mather shares that information with the Royal Society. He writes that Onesimus "had undergone an operation, which had given him something of the smallpox and would forever preserve him from it." Mather adds that "he described the operation to me, and showed me in his arm the scar which it had left upon him."

Mather learns that London's Royal Society has been discussing inoculation, and that a doctor in Constantinople supports the practice. And he knows Onesimus is smart. So he pays attention.

A Brief Rule to guide the Common People of *New-England* how to Order themſelves and theirs in the *Small-Pox* and *Meaſels*.

THE *Small Pox* (whoſe nature and cure the *Meaſels* follow) is a diſeaſe in the blood, endeavouring to recover a new form and ſtate.

2. THIS nature attempts — 1. By Separation of the impure from the pure, thruſting it out from the Veins to the Fleſh. — 2. By driving out the impure from the Fleſh to the Skin.

3. THE firſt Separation is done in the firſt four Days by a Feveriſh boiling (Ebullition) of the Blood, laying down the impurities in the Fleſhy parts which kindly effected the Feveriſh tumult is calmed.

4. THE ſecond Separation from the Fleſh to the Skin, or *Superficies* is done through the reſt of the time of the diſeaſe.

5. THERE are ſeveral Errors in ordering theſe ſick ones in both theſe Opera-

A tions

The first few suggestions in the 1721 edition of a book by the Boston physician Thomas Thacher containing advice for the treatment of smallpox and measles

AN
Hiſtorical ACCOUNT
OF THE
SMALL-POX
INOCULATED
IN
NEW ENGLAND,
Upon all Sorts of Perſons, *Whites*, *Blacks*, and of all Ages and Conſtitutions.

With ſome Account of the Nature of the Infection in the NATURAL and INOCULATED Way, and their different Effects on HUMAN BODIES.

With ſome ſhort DIRECTIONS to the UN-EXPERIENCED in this Method of Practice.

Humbly dedicated to her Royal Highneſs the Princeſs of WALES,
By *Zabdiel Boylſton*, F. R. S.

The Second Edition, Corrected.

LONDON:

Printed for S. CHANDLER, at the Croſs-Keys in the *Poultry*.
M. DCC. XXVI.

Re-Printed at BOSTON in N. E. for S. GERRISH in *Cornhil*, and T. HANCOCK at the Bible and Three Crowns in *Annſtreet*. M. DCC. XXX.

The title page from the 1730 edition of Zabdiel Boylston's account of smallpox inoculations in New England

He finds a doctor, Zabdiel Boylston, who is willing to try this preventive medical treatment in Boston. In May 1721, a ship arrives in Boston Harbor; on board are smallpox victims. When the people from the ship mix with folks in Boston, their germs begin to spread. Dr. Boylston starts inoculating some local citizens, which causes an uproar. James Franklin (Ben's brother), a newspaper publisher, opposes inoculation, as do others. Someone throws a grenade into a window of the house where Cotton Mather lives.

But those who are inoculated are much less likely to die from smallpox.

Onesimus has been named by Mather for an enslaved person in the Bible who flees from his master seeking freedom. Onesimus has a similar dream of freedom. He finds paid work and saves his money until he is able to buy his freedom. Meanwhile, he has helped save the lives of many Bostonians who have been inoculated from smallpox.

Who holds that torch aloft? Why, Linnaeus himself; he has no problem with self-promotion. "Names have the same value on the marketplace of botany as coins have in public affairs," he says. In other words, he expects people across the globe to value the names he gives to plants and animals. And, mostly, they do.

In his great book, *Systema Naturae*, which goes through twelve editions in his lifetime, Linnaeus attempts to name all the known organisms and includes about six thousand plants and more than four thousand animals. Given his system of Latin names, a botanist in Moscow and a botanist in Pennsylvania can exchange information and know they are describing the same plant or animal.

Linnaeus's naming pursuit, known scientifically as *taxonomy*, turns out to be good business. A well-respected London Quaker merchant, Peter Collinson, writes to him, "We are very fond of all branches of Natural History; they sell the best of any books in England." Collinson also sells books to the science-minded in the British colonies in America.

CAROLI LINNÆI
EQUITIS DE STELLA POLARI,
SYSTEMA
NATURÆ
PER
REGNA TRIA NATURÆ,
SECUNDUM
CLASSES, ORDINES,
GENERA, SPECIES,
CUM
CHARACTERIBUS, DIFFERENTIIS,
SYNONYMIS, LOCIS.
TOMUS II.
EDITIO DECIMA, REFORMATA.
Cum Privilegio S:æ R:æ M:tis Sveciæ.
HOLMIÆ,
IMPENSIS DIRECT. LAURENTII SALVII,
1759.

The title page of Linnaeus's Systema Naturae

Science, in much of eighteenth-century Europe, becomes a popular pursuit in middle-class households, often involving family outings to look for remnants of the past. Almost anyone can take part: by digging, by mixing chemicals, by planting a garden, by raising animals, by breeding and crossbreeding, by exploring faraway lands. Scientific societies sprout like weeds. And scientific discoverers become cultural heroes. The data that is getting collected cries out for an organizing genius, which is how Linnaeus sees himself.

An engraving showing Carl Linnaeus in some traditional Saami clothing, which he brought back with him from an expedition to northern Sweden

Picture a grocery store with ketchup next to oranges and ground beef with ice cream. Biological classification was like that before Linnaeus appeared. He brought order to the chaos.

Previously, plants and animals had Latin names that were often five or more words long. Linnaeus devises a two-word Latin naming scheme. The first word, noting genus, is always capitalized; the second word, citing species, is not. Take our name, *Homo sapiens*: *Homo* indicates the genus *Homo*; *sapiens* is our species within the genus. Where do the names come from? Linnaeus finds inspiration wherever he can. *Homo* means "man" in Latin. As Shakespeare wrote in *Henry IV, Part I*, "'homo' is a common name to all men."

Linnaeus expects to name all living things, and he seems to have the zeal and determination to carry this plan through. But he has no idea just how diverse and huge the living world is—or that life evolves. He does understand that he has set a giant task for himself.

"I know no greater man on earth," says the Swiss philosopher Jean-Jacques Rousseau when he describes Linnaeus, who seems to agree. He says of himself, "*Deus creavit, Linnaeus disposuit*," which is Latin for "God created, Linnaeus organized." He calls himself "the Prince of botanists."

The German polymath Johann Wolfgang von Goethe (GER-tuh) fawns after Linnaeus's death: "With the exception of Shakespeare and Spinoza, I know no one among the no longer living who has influenced me more strongly."

Is he overstating things? Not in his time. Goethe (1749–1832), who today is mostly remembered for his poetic works, studies plants intensely. Morphology, the branch of modern biology based on the appearance and analysis of internal and outward structures, traces its lineage to Goethe.

But Linnaeus doesn't deal with life's past or future; he organizes life as it exists in the present, beginning with the familiar kingdoms of plants and animals and working his way to ever-more-specific groups—phylum, class, order, family, genus, *and* species—each nesting in the last like a Russian doll.

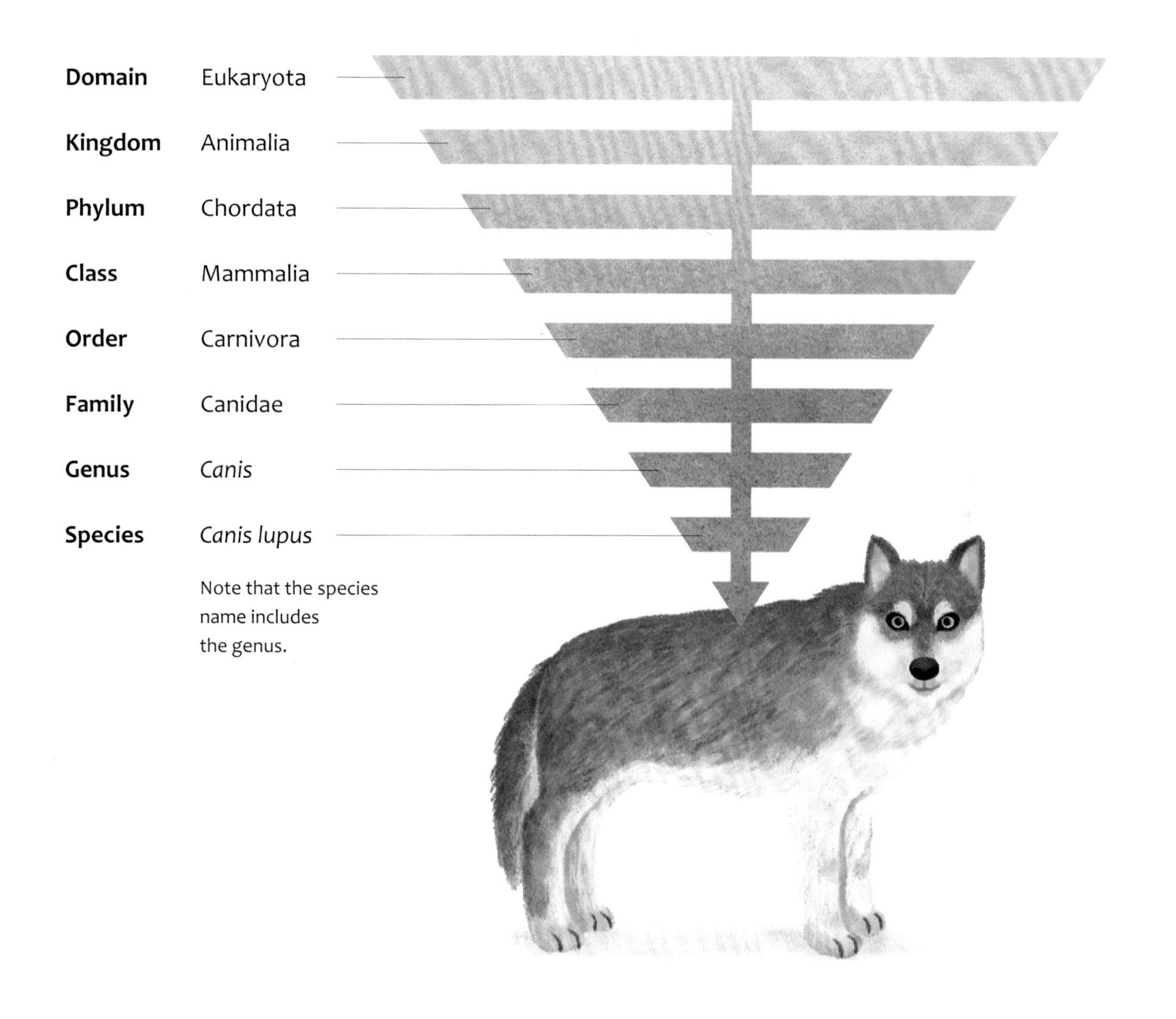

The classification system that Linnaeus begins is still being tweaked and expanded today, with three groups known as domains (Bacteria, Archaea, and Eukaryota) being the broadest categories, groups called kingdoms below domains, and a number of other groupings that get more and more specific, ending with species.

Bacteria are single-celled organisms that can look like spheres, spirals, cork-screws, rods, or commas. They are prokaryotes, which means that their cells lack a nucleus.

Most of the cells in your body are eukaryotes. Plants, animals, fungi, and protists are all made up of eukaryotic cells, which have a nucleus.

Archaea were discovered in 1977 by the American microbiologist Carl Woese. They are also prokaryotes, with cells that lack a nucleus. Archaea are single-celled organ-isms that can live in harsh environments, like vents deep in the ocean that seep heat from the core of the Earth. Their cell walls and membranes are different from those of bacteria.

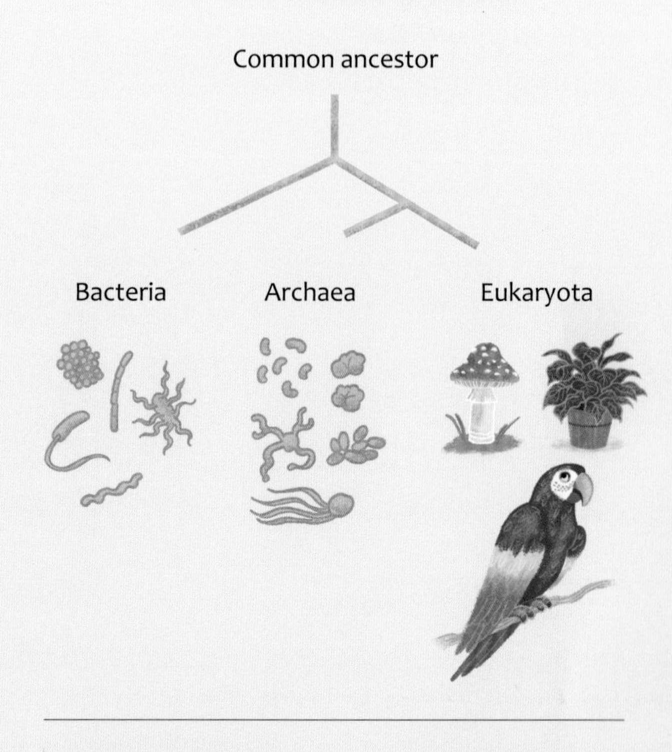

Prokaryotic Cell

No nucleus to contain genetic material

Cell wall

Many prokaryotic cells have an appendage like this flagellum, which helps them move around.

Eukaryotic Cell

Nucleus with genetic material inside

Plasma membrane

If you are interested in time travel, you might consider the Netherlands in the seventeenth century as a destination. The Renaissance, having stretched north, has made it to the Low Countries (the Netherlands and what is today Belgium), bringing business prosperity along with an interest in science and art. Amsterdam becomes a culturally diverse and thriving European economic hub. One reason for this: in 1581, the Dutch rebelled against Philip II of Spain, forming an independent republic that is religiously tolerant (attracting people like Maria Merian); it offers freedom to individuals, to businesses, and to deep thinkers.

A 1685 map of New England and the mid-Atlantic; the inset illustration is a view of New Amsterdam

Artists and scientists emerge: great painters, including Johannes Vermeer and Rembrandt van Rijn (1606–1669); philosophers such as Baruch Spinoza (1632–1677); scientists such as Leeuwenhoek; and scientist-artists such as Merian.

The creative spark found in the Amsterdam of this era is not unfamiliar to Americans today. The Dutch bring a similar spirit to a city they found on the other side of the Atlantic. Today it is called New York, but it once had another name: New Amsterdam.

For a long time, two kingdoms—plant and animal—seem broad enough to hold all of life. Minerals, being nonlife, are handed to geologists.

Given time, though, some bumps arise in what has seemed to be Linnaeus's perfect roadway. In 1866, Ernst Haeckel figures out that single-celled protists don't fit either the plant or animal kingdom. He will devise a three-kingdom system. And more kingdoms are to come. High-powered microscopes will make it clear that microbes are Earth's dominant life-form. ■

Sea anemones, from artwork by Ernst Haeckel

10

A Big-Time Adventurer and a Quiet Scholar

I shall collect plants and fossils, and with the best of instruments make astronomic observations. Yet this is not the main purpose of my journey. I shall endeavor to find out how nature's forces act upon one another, and in what manner the geographic environment exerts its influence on animals and plants. In short, I must find out about the harmony in nature.

—Alexander von Humboldt

I write this letter without knowing where it will find you, but wherever that may be, I am sure it will find you engaged in something instructive for man. . . . Your name is revered among those of the great worthies of the world.

—Thomas Jefferson to Humboldt

LEFT: This 1810 painting portrays Alexander von Humboldt and two companions viewing the Cayambe volcano, in modern-day Ecuador, during Humboldt's expedition to the Andes.

In 1800, there is only one person in Europe as famous as Napoleon, and that is Germany's Baron Friedrich Wilhelm Heinrich Alexander von Humboldt (1769–1859).

An 1806 portrait of Alexander von Humboldt by Friedrich Georg Weitsch, which shows Humboldt working in the Venezuelan jungle

Brilliant, handsome, and athletic, Humboldt is neither warrior nor politician. Rather, he is a scholar and a man of action: a naturalist and polymath who, starting in 1799, treks from the depths of the Amazon jungle to the heights of the Andes Mountains and then writes eloquently about his travels. Just about everyone who reads books reads Humboldt's. His tales inspire a generation of would-be adventurers. American cities name streets and avenues for him. The American essayist Ralph Waldo Emerson calls Humboldt "one of those wonders of the world, like Aristotle." Emerson may be overstating things, but Humboldt is a wonder.

Among his talents is a memory that holds on to the smallest details. He is able to "compare the observations he had made all over the world several decades or thousands of miles apart," writes Andrea Wulf (in a brilliant biography). Humboldt is a scientist at a time when the modern way of examining life is just being born. He writes of his scientific research and his thrilling exploits in *Personal Narrative*, a book with descriptions of South America's dramatic landscapes, of its flora and fauna, and of its peoples—all mostly unknown to Westerners.

Humboldt's observations astound a generation and rightfully so. He has an open mind, and he is eager to learn: when he explores, he considers ideas as well as places. Perhaps his greatest work is his last, a multivolume exploration of the known world and the history of science that he calls *Kosmos*, adopting a word he has found in ancient Greek to consider the wholeness of the universe. And before anyone else seems to have noticed, his close observations of the environment make him aware of human-induced climate change.

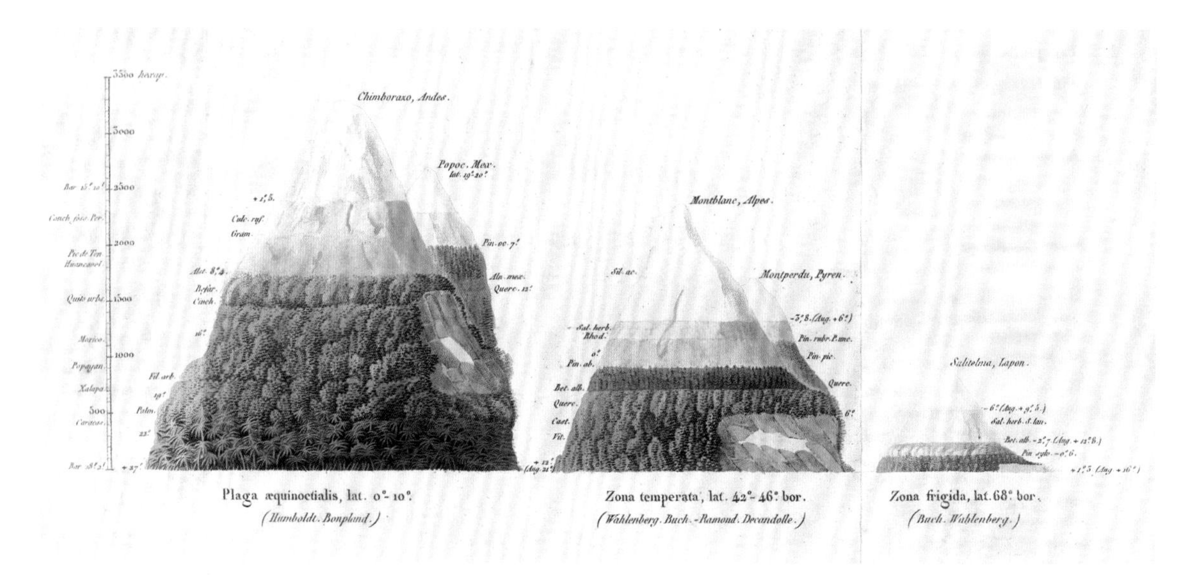

This foldout diagram is from Thomas Jefferson's copy of an 1817 book by Humboldt about the distribution of plants on mountains. The illustration is of the Chimborazo volcano, in Ecuador, along with four other mountains in different temperature zones.

The baron learns about a young scholar, Christian Gottfried Ehrenberg (1795–1876), who has trekked to Egypt, Libya, Lebanon, and the Sinai Peninsula, searching for microscopic specimens of life. On that productive trip, Ehrenberg found specimens of thirty-four thousand animals and forty-six thousand plants, and he also studies fossils of extinct species. All are too small to be studied by the unaided eye.

Humboldt arranges to meet Ehrenberg at the University of Berlin; there he asks Ehrenberg to join him on an expedition sponsored by Russia's ruler, Czar Nicholas I. Ehrenberg is thrilled. Of course he accepts.

The famous explorer and the young scientist are soon exploring Russia, central Asia, and Siberia. They observe, take notes, and gather specimens.

A nineteenth-century engraved portrait of Christian Gottfried Ehrenberg

Ehrenberg and Humboldt discover soil that stays at or below freezing year-round; it will be known as permafrost and is sometimes more than 3,000 feet (914 meters) deep. In the Caspian Sea and in Siberia's Lake Baikal, Ehrenberg finds microorganisms unknown to Western science. After they return home, Ehrenberg begins organizing microscopic life into categories, which no one has done before.

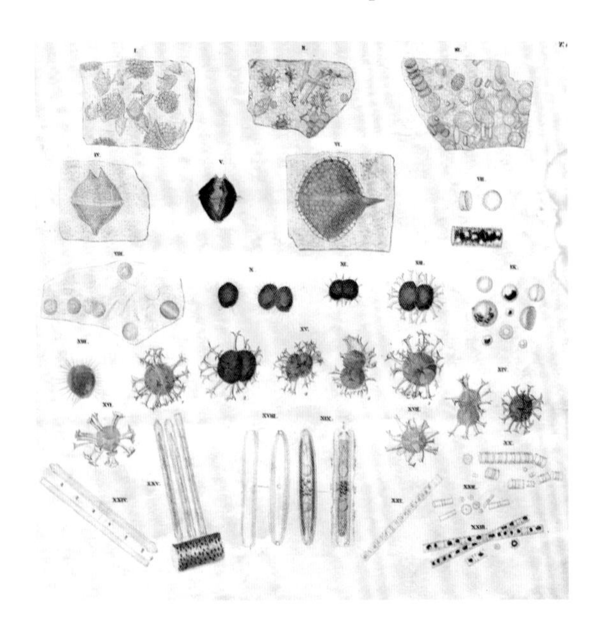

Illustrations of microbes observed by Ehrenberg in the 1830s

What's a microbe? It's a single-celled organism defined by its tiny size. In a human, it can be bacteria or archaea or fungi. Sometimes microbes can be harmful and even deadly for humans, but mostly they are essential to our existence. For example, they help keep our digestive system working. Microbes also play a recycling role on our planet by breaking down dead plants and animals and turning them into organic matter. We couldn't exist without microbes.

In contrast, when the great Swede Linnaeus peeked into the microbial world, he put all microbes into a category he called *chaos*, saying that microbes are so tiny that no one would ever be able to study them with precision.

But that is exactly what Ehrenberg does. He builds a wood-frame microscope so he can look closely at the tiny aquatic creatures that are being called infusoria. He observes protozoa and one-celled algae and notes that each creature is full of life. Ehrenberg says that the matter that makes up life (organic matter) "reaches further back in the history of the earth than had hitherto been suspected."

When he finds fossils in Earth's rock layers, he studies them with his microscope and realizes that some rock masses, like chalk cliffs, are made of the shells of life-forms that have deteriorated and piled up over great lengths of time. He is rediscovering Steno's law of superposition—that layers of rock are in fact layers of time, with the oldest layers at the bottom.

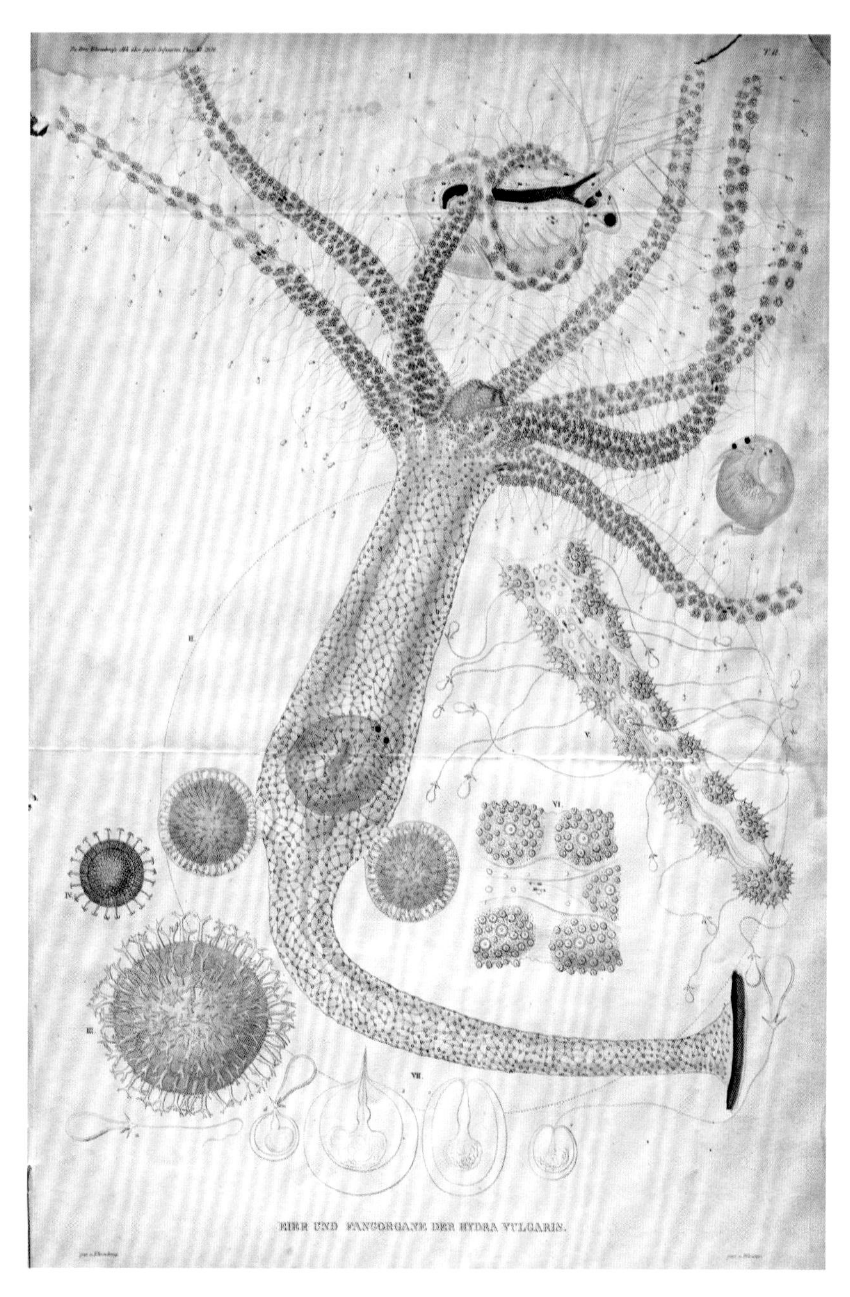

An illustration from Ehrenberg's 1837 book on microbes of a Hydra vulgaris, *a freshwater creature that is little more than an inch (30 millimeters) long at most*

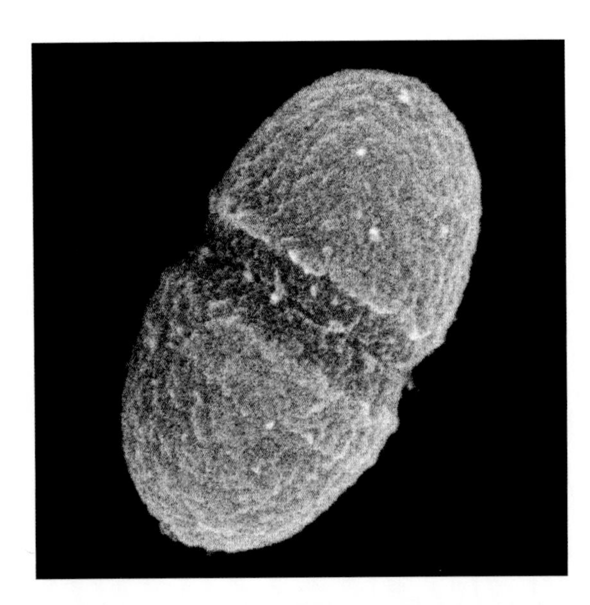

The bacterium Enterococcus faecalis, *which is found in the human gut*

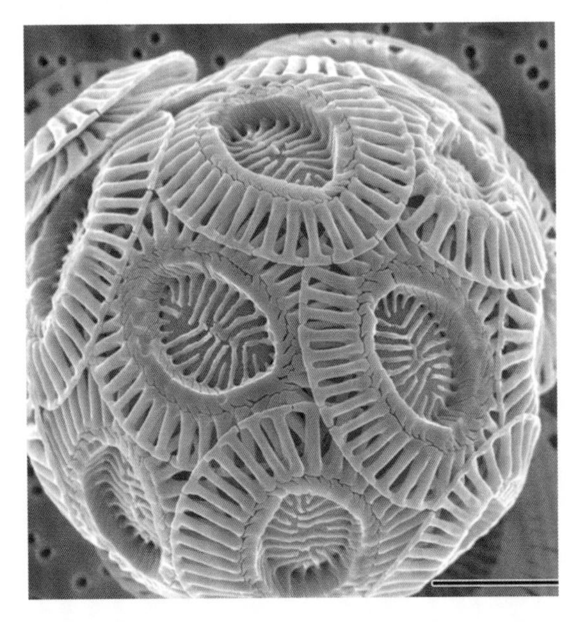

A single-celled marine creature called a coccolithophore as seen through an electron microscope

Ehrenberg realizes that a huge amount of time must pass for a chalk cliff to take shape; that means Earth's history must have provided that time. This insight, also shared by others, will gradually change the prevailing ideas on the universe: its age and its origin.

Chalk is created by tiny marine creatures who secrete calcium carbonate to make shells, and that carbonate gets layered bit by bit by bit. Ehrenberg studies the calcium carbonate of the chalk cliffs and finds it is a chemical compound made of three main elements: carbon, oxygen, and calcium. Those are the core ingredients of eggshells, snail shells, and pearls. He discovers fossils of microbes and studies them. His work will help establish the science of micro-paleontology. Ehrenberg is soon linking biology and geology and history as no one before him has done.

Ehrenberg is paying attention to cells and life-forms that most others ignore. He writes his doctoral dissertation on fungi, including yeasts and molds. In it he describes 250 species of fungi found in the Berlin area. He notes that some fungi are single-celled (called unicellular), and some are multicelled. This is new knowledge.

Most scientists of the time believe that fungi appear spontaneously, for no known reason.

The Cell: Life's Essential Particle

Theodor Schwann (1810–1882), who studies at the University of Berlin while Ehrenberg is a professor there, spends hours peering into microscopes. In 1838, he and a colleague, Matthias Schleiden (1804–1881), compare what each has seen through his lens. They use Robert Hooke's word *cell* to describe a structure they realize is common in all life-forms. Schleiden can see that plant tissue is made of cells and that each plant cell has a nucleus. Schwann has found the same thing in animals—they too are composed of cells, and each has a nucleus.

Matthias Schleiden

Schwann and Schleiden come to realize that biology's basic unit is the cell. They aren't the first to see cells. Several biological explorers have already grasped the essential role of the cell and that it can exist both as an independent living entity and as a building block in a greater whole. But the two men are credited with developing cell theory, or the idea that cells are the building blocks of life.

Schwann figures out that complex multicelled organisms (like us) develop from a single fertilized egg cell. These two agree that something cannot come from nothing; so, for them, there can be no spontaneous generation. A French scientist-politician, François Vincent Raspail, who also agrees, uses a Latin phrase: "*omnis cellula e cellula*," which

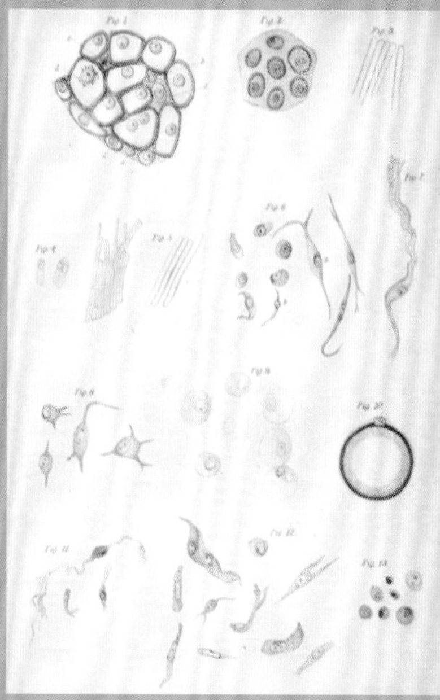

Illustrations of cells (most from a fetal pig) from Schwann's book Microscopical Researches into the Accordance in the Structure and Growth of Animals and Plants, *with contributions from Schleiden*

means "each cell comes from another cell." Later, an important doctor, Rudolf Virchow (1821–1902), will get credit for that phrase, which becomes basic to cell theory, a foundation of modern biology.

The nineteenth century's thinkers and experimenters lay a base for the microbiology to come. Matthias Schleiden says that anyone "who expects to become a botanist or a zoologist without using the microscope . . . is as great a fool as he who wishes to study the heavens without the telescope." These pioneering biologists are not fools.

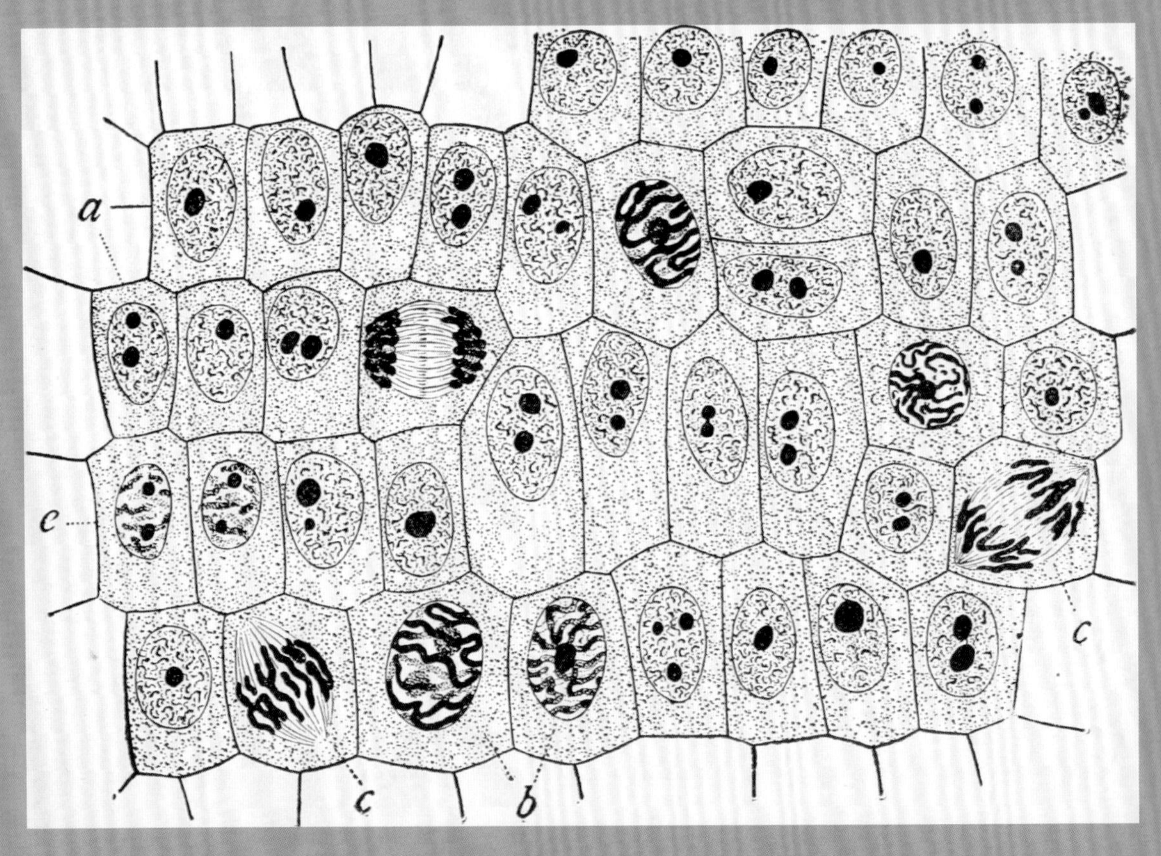

A drawing of cells in the roots of an onion, seen at a microscopic level

Ehrenberg finds the reason: molds and mushrooms reproduce sexually, and fungi come from spores that have reproductive structures somewhat like those of seeds (although without stored food). Mushrooms are the "flowers" of some fungi. Ehrenberg becomes the world's expert in these fields.

Meanwhile, microscopes are getting better: brass replaces wood for the frames, the content of the glass lenses changes, and the distance between lenses becomes fixed. Ehrenberg gets magnifications of about three hundred times. He draws what he sees and publishes his drawings; those drawings, like Hooke's, are art as well as science.

How big is a cell? Cells come in different sizes, even those within the human body. Liver cells are average-sized human cells. About ten thousand liver cells, end to end, would cover about 20 inches (50 centimeters). Some nerve cells are giants, stretching from the base of the spine all the way to the tip of the big toe.

If you want to think small, consider red blood cells—they are ten times smaller than most human cells, roughly 0.725 inches (18 millimeters). It would take seventy thousand red blood cells to stretch 20 inches (50 centimeters).

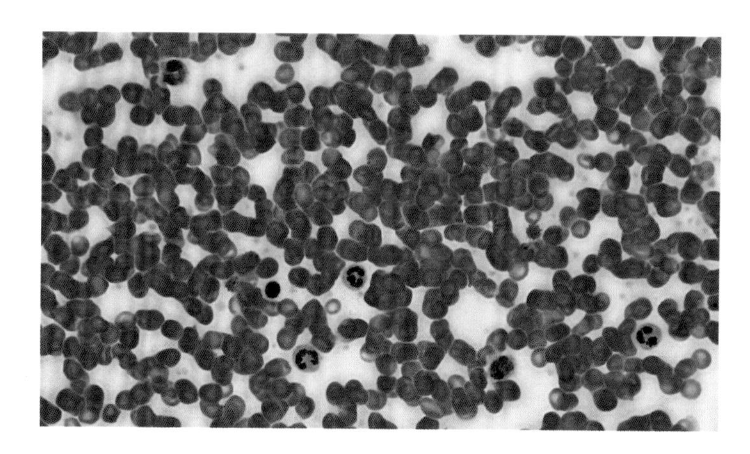

Human blood cells at 400× magnification

We now realize that microbes are not all the same. Some are bacteria, some are archaea, some are eukaryotes, some are viruses. Microbes are not a family—the word *microbe* just describes their size, which is very small. Today's electron microscopes let us see the microbial world in detail, so we are aware of bacteria, archaea, fungi, algae, protozoa, and viruses. These creatures share their small size but often little else. In 1838, when Ehrenberg writes that microbes "are very inferior in individual energy to lions and

Color plate of a variety of organisms from Ehrenberg's 1837 book on microbes

elephants, but in their united influences they are far more important than all the animals," he doesn't realize how important they really are.

But Ehrenberg was right: when you consider them all together, microbes dominate our Earth. Hardly anyone understands that in the nineteenth century, although a few scholars do begin to pay attention to microbes. One of them is an English naturalist named Charles Darwin (1809–1882). He will sail around the world on a scientific mission when he is in his twenties and send infusoria (microbes) he finds on the ship's deck to Christian Ehrenberg.

Darwin will take Humboldt's *Personal Narrative* with him on his voyage, calling the book a "rare union of poetry with science." When he writes a fan letter to Humboldt, he gets an answer. Young Darwin is thrilled and reads Humboldt's letter aloud to his friends over and again. ■

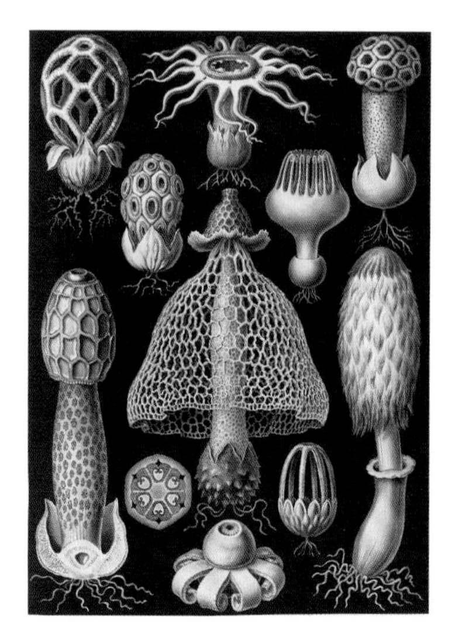

A variety of fungi, from Ernst Haeckel's 1904 book, Kunstformen der Natur (Art Forms in Nature)

11

Three Icons; One Is Pasteurized

Bacteria represent the great success story of life's pathway. . . . The number of Escherichia coli *cells in the gut of each human being exceeds the number of humans that has ever lived on this planet.*

—Stephen Jay Gould

We've tried ignorance for a thousand years. It's time we try education.

—Dr. Joycelyn Elders

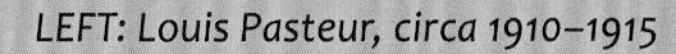

LEFT: Louis Pasteur, circa 1910–1915

In the mid-nineteenth century, most people are unaware that creatures too small to be seen by human eyes are in, on, and all around them. These "animalcules," discovered by Leeuwenhoek, are known principally to the scholarly and the inquisitive. Hardly anyone else understands that the world of tiny creatures is essential to human life.

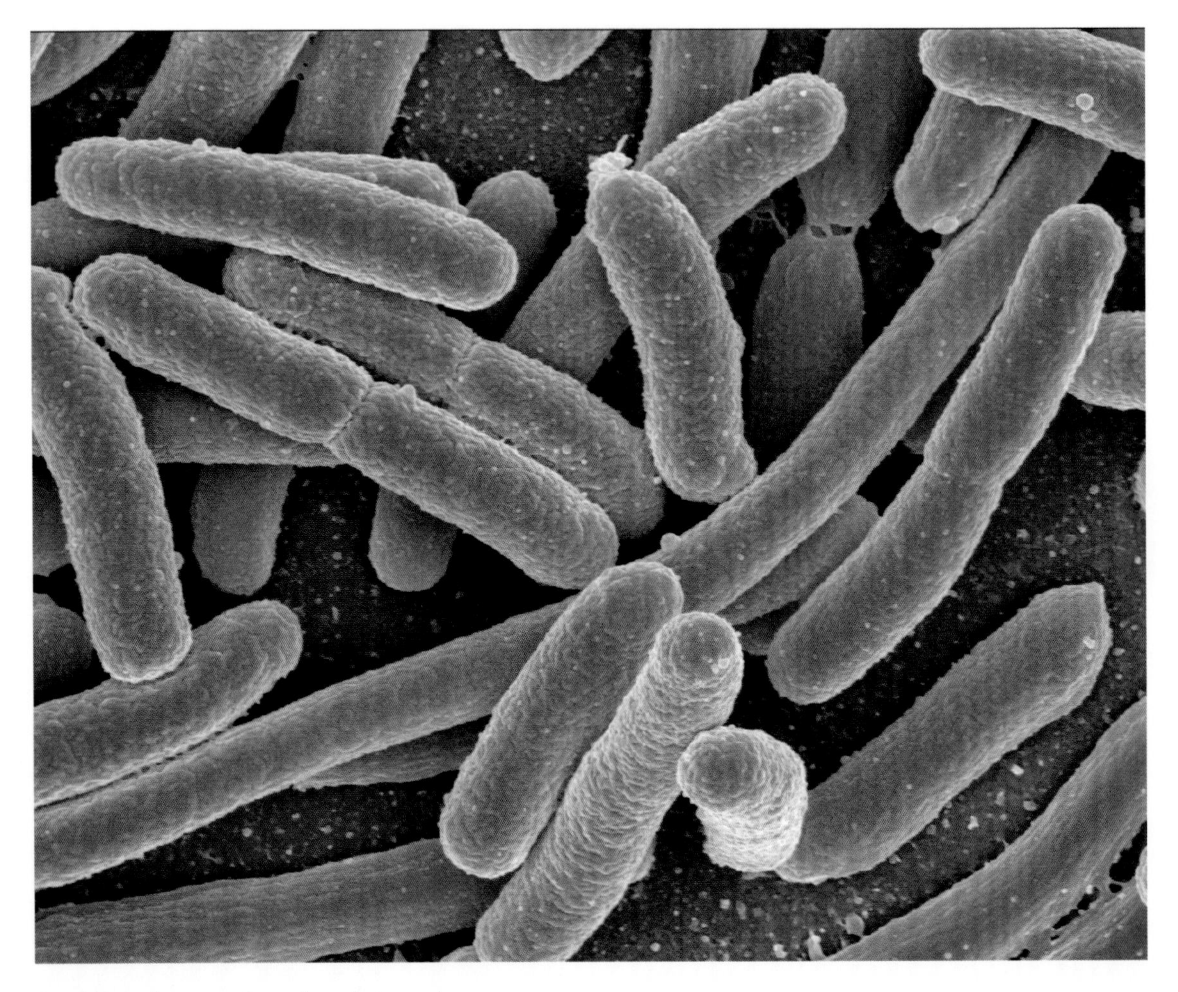

E. coli *bacteria as seen through an electron microscope*

And then a trio of brilliant scientists comes into prominence: Cohn, Koch, and Pasteur. They are all familiar with Ehrenberg's work.

Ferdinand Cohn

One member of that nineteenth-century trio of biologists, Ferdinand Cohn (1828–1898), is today known as a father of modern bacteriology. Cohn is Jewish in a highly anti-Semitic time and place, which means that although he studies at the university in his hometown of Breslau (then in Germany, now in Poland), he is not allowed to get his doctorate degree there. Cohn wants to learn, so he leaves home and enrolls at the University of Berlin. He will have a doctorate in botany by the time he is nineteen.

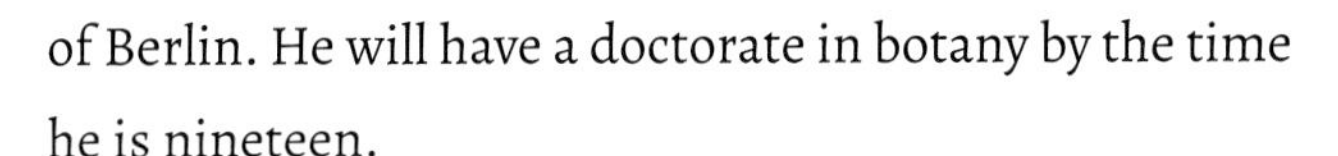

After graduation, he goes back to Breslau and sets up his own lab at home, and soon he becomes a lecturer, and then professor, at the university that once snubbed him. Cohn will spend the rest of his life studying algae, protozoa, fungi, and bacteria. He figures out that drinking water, if it contains certain microbes, can be a source of disease; going further, he begins to understand the role of some bacteria in infectious diseases.

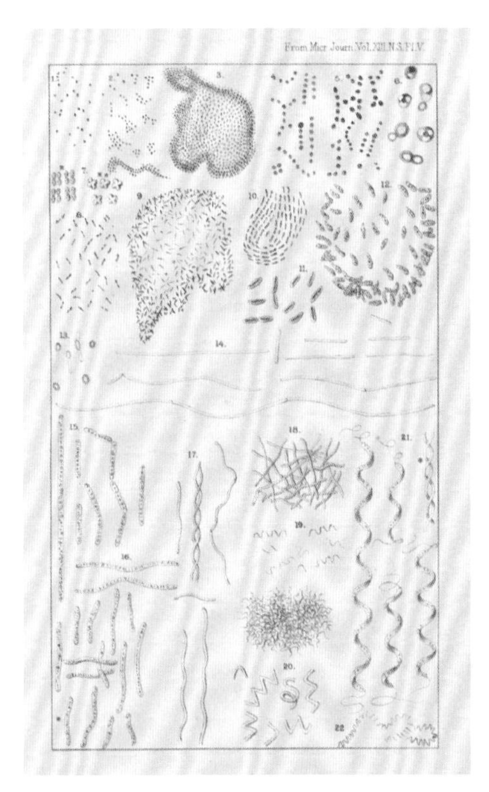

A page from Cohn's 1881 book, Bacteria: The Smallest of Living Organisms, *showing his four categories of bacteria: rods, threads, spheres, and spirals*

Cohn starts classifying bacteria, something that most others think impossible to do. He sorts bacteria into four categories depending on shape. Some look like rods, some like threads, and others are spherical or spiral. His categorization system remains in use today.

Cohn's lab becomes one of the best in the world. When Robert Koch (1843–1910), a young medical doctor in the town of Posen, learns what Cohn has been doing, he makes contact, and Cohn soon becomes a mentor to Koch.

Algae

Protozoa

Fungi

Algae are organisms that use photosynthesis (just as plants do) to provide nourishment. Seaweed and pond scum are among algae that can do that.

Protozoa are single-celled microscopic animals that have a nucleus, which is encased in a membrane within the cell and contains the cell's genetic materials. (Protozoan cells can even have more than one nucleus.) Amoebas, flagellates, and sporozoans are protozoans. Some, like the protozoans that cause the disease malaria, are harmful to humans. All belong to the kingdom Protista.

Fungi (the singular is *fungus*) produce spores that feed on organic matter. Mushrooms, mold, and yeast are fungi.

Koch, who surprised his parents when he taught himself to read at age five, has set up a lab in his small apartment, where he grows bacterial cultures, sees patients, and lives with his wife. In 1871, on his twenty-eighth birthday, she gives him a microscope as a present.

Koch believes that microbes play a role in disease, and that each disease must have its own microbe. He sets out to see if he is right. But how do you isolate a microbe? The microbes swimming on his slides are all mixed up, and he can't seem to separate them.

Then Julius Richard Petri, Koch's assistant, fills a circular glass dish with agar, a jelly-like substance. That surface isolates cultures of bacteria. In Petri's dish, bacteria that are different from each other don't mix, which is very helpful to scientists who want to study them. "Petri" dishes soon become a standard in scientific laboratories.

Meanwhile, France and Germany are fighting each other in the Franco-Prussian War (1870–1871); Koch is called to the battlefield as a doctor. There he sees a lot of disease that doesn't come from war injuries, which surprises him. Frustratingly, he can't do much for most of his patients, who are living in crowded army barracks. It becomes clear to him that some diseases are passed from human to human and animal to animal and that those barracks are breeding grounds. But why? And how?

He studies anthrax, a disease that is killing sheep and cows in Europe. He realizes that it seems to go dormant and then reappear. Why? No one knows.

An 1887 portrait of Robert Koch

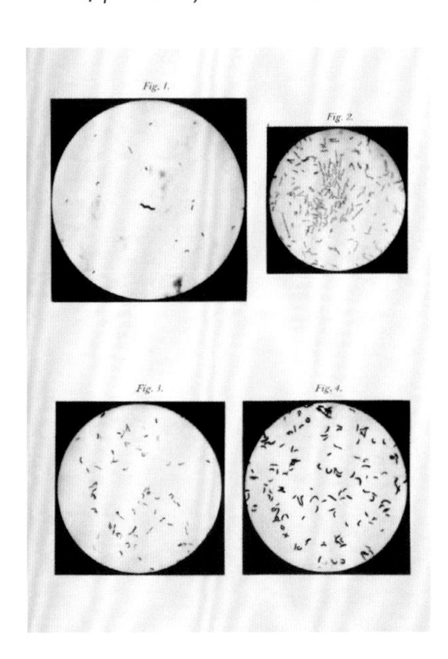

Cultures from an 1893 book by Julius Petri

In 1876, Koch writes to Ferdinand Cohn in Breslau:

> *Honored Professor! I have found your work on bacteria . . . very exciting. I have been working for some time on the contagion of anthrax. After many futile attempts I have finally succeeded in discovering the complete life cycle of* Bacillus anthracis.

Koch, who often spends hours peering into his microscope, has found and isolated a bacterium that seems to carry anthrax. Koch studies the rodlike anthrax structures. He does experiments. In his lab, he puts anthrax cells in inhospitable conditions and watches their spores go dormant. When he improves their conditions as part of his experiment, the spores awaken.

Robert Koch (front row, fourth from left) with the students taking his first postgraduate course in bacteriology, in Berlin, 1891

He takes liquid from the spleens of animals that have died of anthrax, as he believes that liquid will help him understand how anthrax works—and he is right. When he injects the diseased body liquid into mice, the mice get anthrax and die. Koch is making important discoveries about how disease spreads.

Koch shares his work with Ferdinand Cohn, who publishes Koch's findings in a scientific journal, where it gets attention. Koch is soon a major player in the emerging field of microbiology; he travels to India, South Africa, Egypt, and

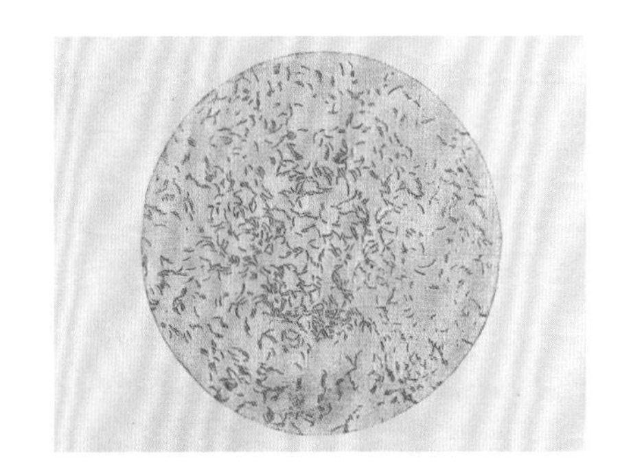

A culture of the bacteria that cause tuberculosis as seen under a microscope, from an 1891 pamphlet about Koch's discovery

Koch working in his laboratory. In 1882, following his work on anthrax, after six months of intense investigation, he identified the bacterium that causes tuberculosis.

This depiction of Koch was part of a series called "Benefactors of Humanity" by the French chocolate brand Chocolat Guérin-Boutron.

Java, studying disease. He discovers the bacteria that cause tuberculosis and cholera. The small-town doctor is soon world-famous.

Robert Koch establishes four postulates, or rules, that link microbes and disease; they are still seen as foundation stones in the field of microbiology and disease:

1. A microorganism must be found in abundance in all organisms that carry the disease, but not in healthy ones.
2. The microorganism must be able to be isolated and grown in pure culture.
3. The cultured microorganism should usually cause the disease when introduced into a healthy organism.
4. When a microorganism is isolated from the experimental host, it must not change.

It is still hard during this time for some people to accept that organisms too small to be seen by human eyes can kill people. Understanding that will lead to the development of the germ theory of disease—the idea that certain sicknesses are caused by very tiny microorganisms that invade the body. It takes a scientist with a knack for publicity to make the general public aware of microbes, of their often devastating power, and of the need for cleanliness in order to control them. Louis Pasteur (1822–1895) is that scientist; he makes people listen.

An 1892 etching of Louis Pasteur in his lab

Pasteur is the son of a leather tanner in Arbois, France. He is a talented portraitist and will have a lifelong interest in art, but his father discourages him from pursuing art as a career. A gifted and prizewinning student, he is sent to a prestigious high school in Paris. He arrives there at age fifteen but is soon so homesick that his father has to come to Paris to bring him home. There he helps his father with leather tanning (a process that depends on microbes, although no one then realizes it).

After a few years home in Arbois, and studying at a nearby boarding school, Louis Pasteur is ready to move back to a school in Paris, the École Normale Supérieure, though the first time he tries, he fails the entrance exam. Finally, Pasteur finds his passion, which is studying diseases and figuring out how to understand them and control them.

He becomes a scholarly workaholic. When he boils milk and wine, he finds that boiling kills disease-causing microbes that they may carry. That process of boiling will be named pasteurization. He is also a showman and speechmaker who makes people aware of public health issues and even conducts a public trial of a vaccine for sheep. The public is soon awed by this young scientist who happens to have a genius for communicating. He not only gets results; he also takes time to make sure people understand what he has done. Pasteur, who becomes a public celebrity, makes life science popular and understandable.

Pasteur also makes people aware of the importance of antiseptics and cleanliness in preventing disease. So does Koch. Together, they urge the sterilization of surgical instruments and promote simple handwashing. Before this, doctors and the general public had no idea that cleanliness was linked to health. Doctors with bloody hands often went from patient to patient without washing, which spreads disease and infection.

Pasteur makes it clear that cells come from other cells and that new cells don't just pop up out of nowhere. Cells come from cell division.

Pasteur conducted several experiments, one of which is diagrammed here, that refuted the idea of spontaneous generation.

This 1920 postcard shows people working in a laboratory at the Pasteur Institute, in Paris.

If you understand that, you know that spontaneous generation cannot happen. Life can't appear on its own. Pasteur performs an experiment that has been done before: he boils meat broth and puts it in sealed glass flasks that keep out microbes. No worms appear in the sealed containers. People pay attention when Pasteur says or does something. Reporters seek him out. So when he says that life cannot come from nonlife, people trust him. Finally, the theory of spontaneous generation is overthrown.

Pasteur also understands that while some microbes are disease carriers, others are essential to life. He grasps the complexity and importance of the microbial world long before most others do.

In 1887, he founds the Institut Pasteur (Pasteur Institute) in Paris. The institute attracts research scientists from around the world, who will find a cure for diphtheria, develop a rabies vaccine, and do major work in genetics and microbiology. ■

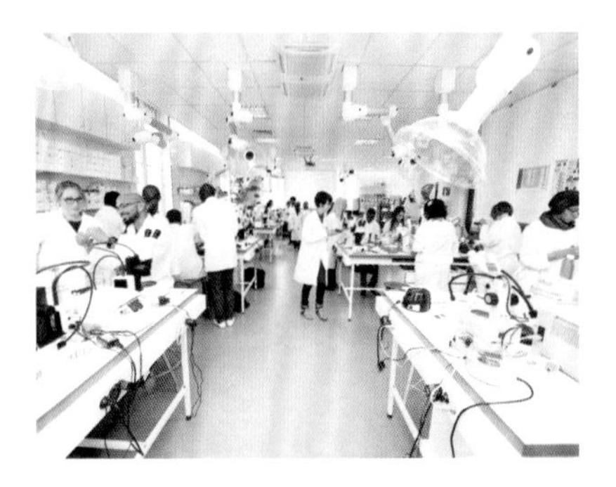

The Pasteur Institute in 2016

Today the Institut Pasteur is a world-renowned center for medical and biological research, with a staff of more than 2,800 and a global network of institutes devoted to medical and biological research, especially in developing countries.

FIG. 1.
FIG. 2.

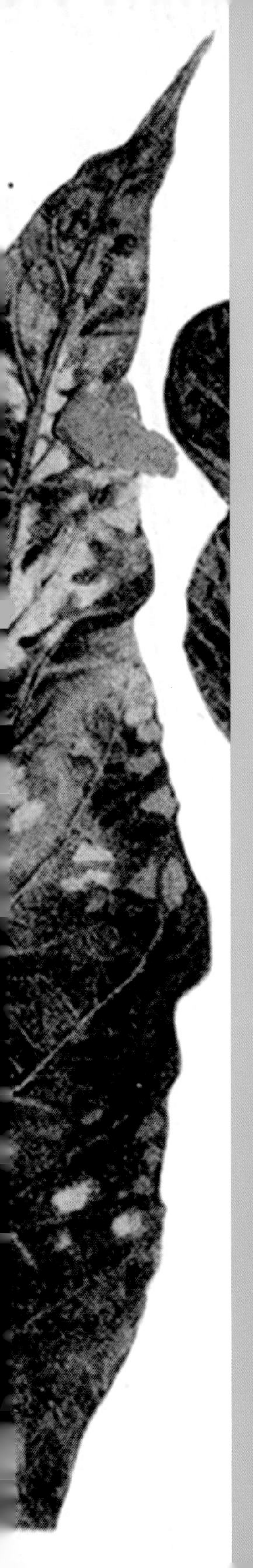

12

A Tobacco Disease in Holland Baffles the Experts

All that we can say for sure is that, because viruses as they now exist are totally parasitic, there must have been cells before there were viruses. There is no way in which we can imagine a virus coming first and later growing into a cell.

—Freeman Dyson

There is a tide in the affairs of men,
Which, taken at the flood, leads on to fortune;
Omitted, all the voyage of their life
Is bound in shallows and in miseries.
On such a full sea are we now afloat.

—William Shakespeare

LEFT: This colorized photograph taken in 1898 by M. W. Beijerinck shows tobacco leaves infected with tobacco mosaic disease.

In the early 1600s, the Dutch establish a North American colony they name New Netherland, which encompasses parts of several future states, including New York and New Jersey. New Netherland has an active fur trade with local Indigenous nations, in particular the Mohawk, but the relationship often turns hostile. The Native Americans and the Dutch have clashing views of land rights. Indigenous peoples tend to view themselves as stewards of the land rather than owners of it, though different tribes do sometimes battle with each other for rights over parts of their territory, in particular hunting grounds. But the European concept of personal land ownership is not the practice of Indigenous nations. Dutch colonists seize land and call it their own.

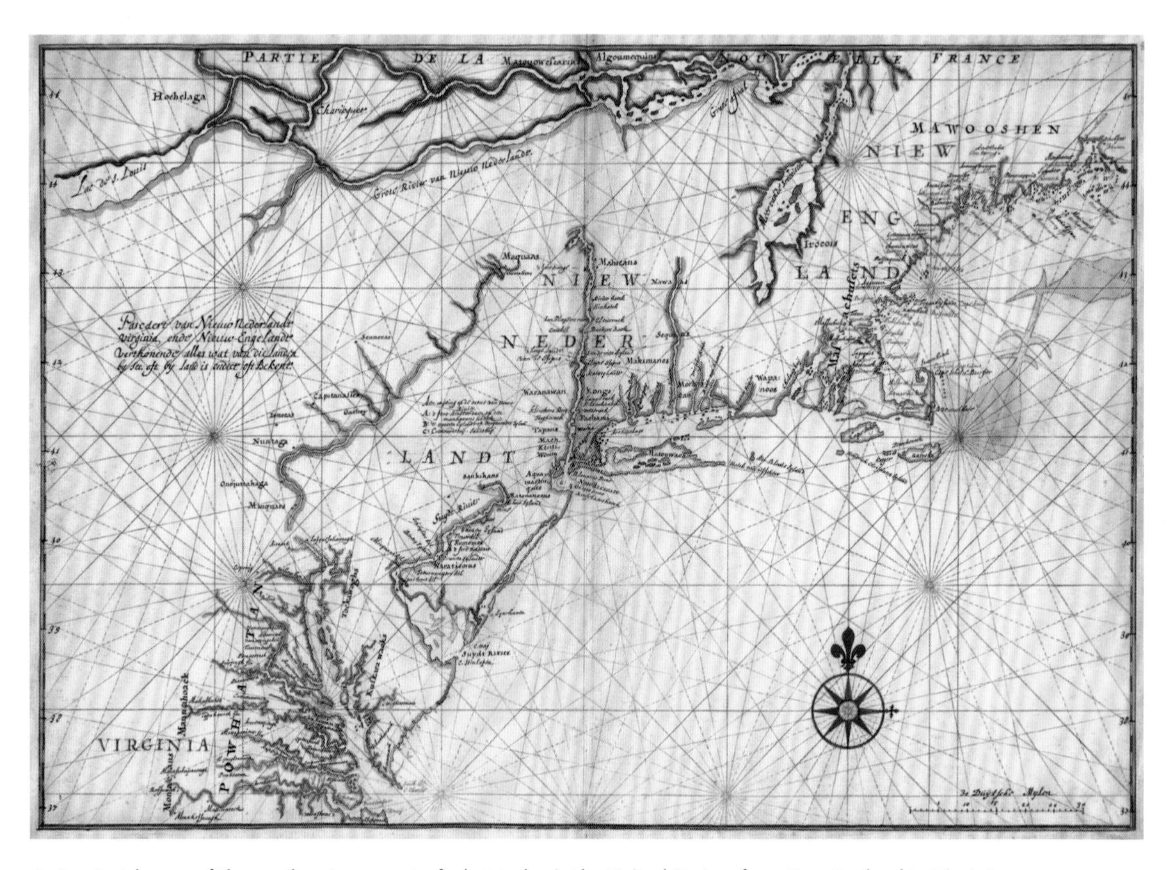

A 1639 Dutch map of the northeastern coast of what today is the United States, from New England to Virginia

Among other things, the Dutch use the land to establish more than two dozen tobacco plantations.

The Dutch begin shipping big tobacco leaves from America to their home across the ocean. In the Netherlands, the tobacco leaves are processed, chopped, put in long clay pipes, and smoked. Dutch farmers soon start growing their own tobacco crops in Holland.

People who are daring or fashionable or just curious begin puffing tobacco. It becomes the rage in much of Europe; some say there are health benefits to smoking. They are wrong, but no one knows that for sure. Tobacco profits turn into gold in the Dutch treasury.

An illustration from a 1622 book purportedly depicting Indigenous people planting and harvesting tobacco. The book saw some virtue in tobacco but warned against its overuse.

Then something awful happens: a blotchy, mosaic-like pattern of live and dead tissue appears on leaves in the Dutch tobacco fields. No one wants to smoke those leaves. By the mid-nineteenth century, whatever is infecting the crop is destroying the livelihood of Dutch tobacco farmers and merchants.

Martinus Beijerinck

One of Amsterdam's bankrupt tobacco merchants is named Beijerinck (say "BY-er-ink" and you'll be close; there is no exact English equivalent of some Dutch vowel sounds). His son, Martinus Willem Beijerinck (1851–1931), happens to be a biologist. Young Beijerinck is eager to find the villain behind the tobacco outbreak. This is an agricultural calamity that is harming the Dutch economy, and it commands attention.

Because of Pasteur, bacteria are now linked to many diseases; they seem plausible suspects in this outbreak. Can the culprit be identified by looking through a lens?

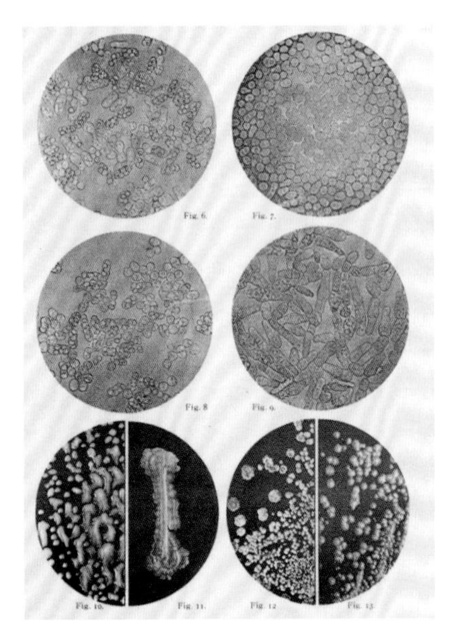

Beijerinck's microscopic observations of yeast and bacteria and their mutations

Microscopes are the technological wonders of this age. A French microbiologist invents a filter with pores smaller than a bacterium. When a solution containing bacteria goes through the filter, the bacteria are separated out; then the latest magnifying microscopes bring them into view. Still, as powerful as these new devices are, there's a limit to what can be seen through them. And whatever it is that Beijerinck is looking for, it is too small to be seen through the microscopes of the day.

Chemist Adolph Mayer (1843–1942), a young professor at the agricultural school in the small town of Wageningen, is among those tackling what is now being called tobacco mosaic disease (TMD). Mayer realizes that the infection is being transferred from plant to plant in the same way that some bacterial infections are transferred from person to person.

He filters bacteria from the sap of infected plants. Then he takes that bacteria and injects it into healthy plants. The plants do *not* get TMD. Mayer is baffled. In a paper written in German in 1886, Mayer says that the mosaic disease "is bacterial, but that the infectious forms have not yet been isolated." He is wrong: the TMD culprit is not bacterial.

Adolf Mayer in an 1875 photograph

In Russia, biologist Dmitri Ivanovsky (1864–1920) does a similar experiment, but he goes further: he filters out the bacteria, and instead of injecting the bacteria into healthy plants, he injects the remaining bacteria-free *sap* into healthy plants. They get TMD. How can that be? The bacteria have been removed.

Ivanovsky is still convinced the infectious agent is a bacterium, but he is wrong.

Beijerinck, who is now a colleague of Mayer's in Wageningen, guesses that whatever is causing the infection may not be a bacterium, but another entity, *one so small it passes right through the filters that trap bacteria*. He runs an experiment like Ivanovsky's in which he filters sap, removes the bacteria, and injects the bacteria-free liquid into a healthy plant; it soon has TMD. Something in that filtered liquid is multiplying in the cells of the tobacco plant. Whatever it is, the infectious agent will not grow on the Petri dishes now being used to cultivate bacteria. The infectious agent exists only in cells that are dividing. Beijerinck figures out that whatever he is looking for needs a host cell in order to reproduce.

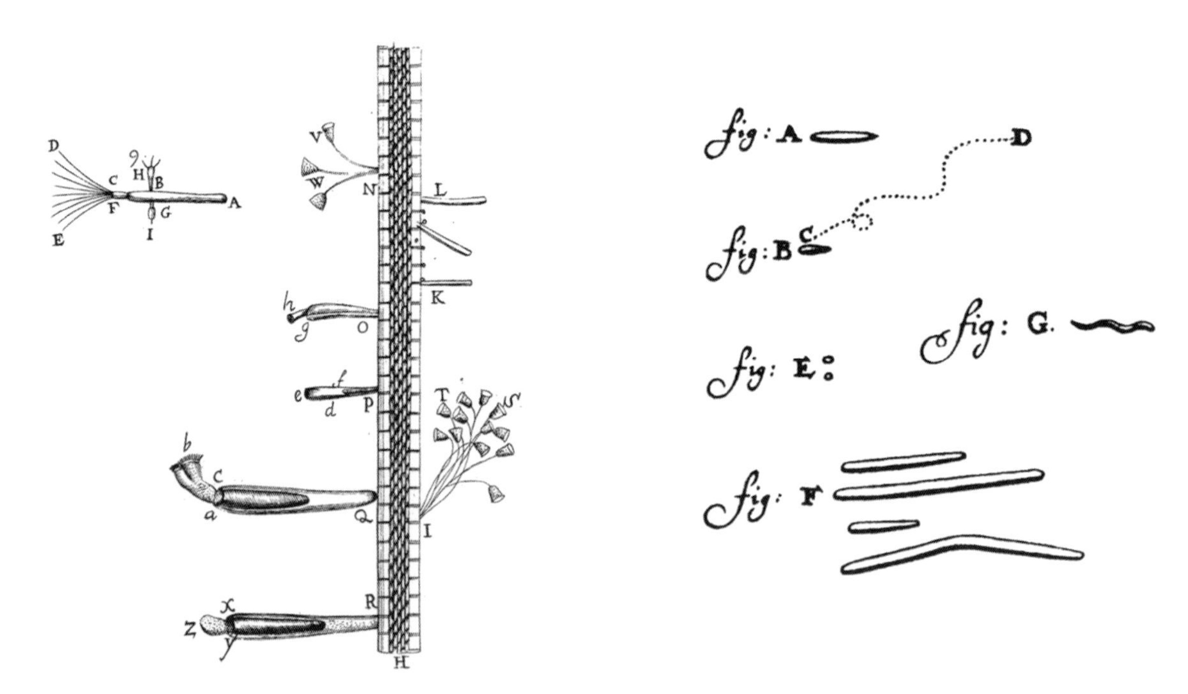

Drawings of organisms observed by Leeuwenhoek under a microscope: the ones on the left came from Dutch canal water where weeds were growing; the ones on the right came from his mouth.

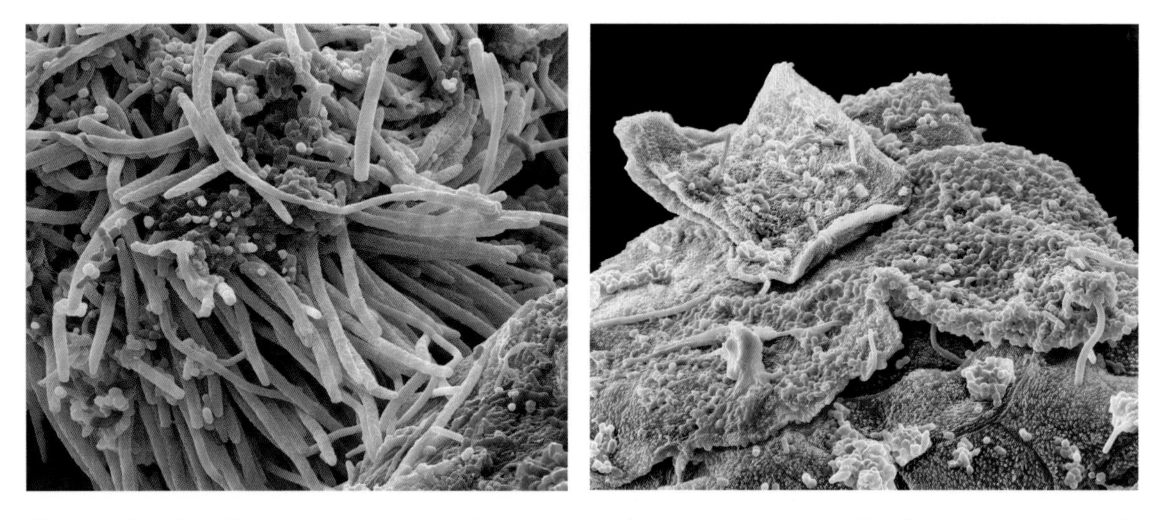

Microorganisms found in someone's mouth. You might want to check what's inside your mouth under a microscope (or not!).

He realizes it must be an astonishingly tiny particle. He calls the unknown particle a virus. The name, which he comes up with in 1898, is derived from the Latin *virulentus*, meaning "poisonous slime." He figures out that these viruses must be at least a hundred times smaller than bacteria. Which means one drop of water can hold millions of virus particles.

That makes them much too small to be seen using *any* optical microscope. And all the microscopes of the day are optical, meaning they work by enlarging an image. But there is a limit to how large a microscope and its visual enlargement can be. A new kind of microscope will need to be invented to see objects that are very small.

That will happen—electron microscopes will be developed—but not in time for Beijerinck to study what they show. The first of them, built in the 1930s, magnifies about four hundred times. (Today some are capable of two million times magnification.)

In 1935, which is four years after Beijerinck's death, the viral agent he discovered is formally named tobacco mosaic virus, which helps scientists understand, though not solve, the problem. ■

Looking Ahead

At the beginning of this book, science was languishing in the long shadow cast by the brilliance of the ancient Greeks and Romans. Because human dissections were discouraged, doctors didn't understand the layout of organs in the human body. They didn't know that our hearts pump blood. Scientists relied on the naked eye, or theories in their head, since there were no microscopes to guide them further.

An 1828 cartoon of a woman dropping her teacup when she sees the gross-looking magnified contents of drinking water

By the Renaissance, the printing press had made books widely available, and translations of the Bible into spoken languages boosted literacy across Europe. In their models, scientists moved the Earth from the center of the universe and realized its proper place as a satellite of the sun. And microscopes opened up what had been a hidden microbial world teeming with abundant life.

There is still much to be learned for scientists in the years after the Renaissance. Pandemics run unchecked, and doctors often practice archaic and dangerous treatments, including "bleeding" patients. While viruses have been discovered, there is little understanding of their role in disease. Women remain relegated to the status of second-class citizens.

But scientists on the horizon will push new boundaries. As Charles Darwin puts it, "A man who dares to waste one hour of time has not discovered the value of life."

For Further Reading

This is a book about life and about the process we human beings went through in order to better understand how life works. It wasn't easy for us humans to figure life out, and sometimes it wasn't easy for me to write about it.

I began by reading stories, true stories, of life science's explorers. I read books, a whole lot of them; I also read articles in magazines and newspapers. I looked for information on the web. I visited several scientists and asked questions.

Here are some notes on books that I found especially helpful, that informed this work, and that are recommended for readers who want to delve more deeply into particular subjects:

Arikha, Noga. *Passions and Tempers: A History of the Humours.* New York: HarperCollins, 2007. This is a history of Western medicine, and a good one. It begins in ancient Greece before Socrates and goes on to modern times. The pages on Renaissance and early modern thinkers hold insights that will help you understand how we have blended reasoning from the past with the findings of modern science.

Bacon, Francis. *Complete Essays.* New York: Dover, 2008.

Bryson, Bill. *Seeing Further: The Story of Science, Discovery, and the Genius of the Royal Society.* New York: HarperCollins, 2010. Bill Bryson's official history of the Royal Society.

Bynum, William. *A Little History of Science.* New Haven, CT: Yale University Press, 2012. This small book attempts to deliver an overview of Western science (physics as well as biology). It's short and easy to read.

Carroll, Sean B. *The Story of Life: Great Discoveries in Biology.* New York: W. W. Norton, 2019. Carroll adds enthusiasm and insights to some of biology's best stories.

Dobell, Clifford. *Antony van Leeuwenhoek and His "Little Animals": Being Some Account of the Father of Protozoology & Bacteriology and His Multifarious Discoveries in These Disciplines.* New York: Dover, 1960. This is a wonderful book about a man, fascinated by microscopes, who built his own, which was better than the professional ones. Then he used it to see remarkable things that he shared with the world.

Frankopan, Peter. *The Silk Roads: A New History of the World.* New York: Vintage, 2017. Frankopan expands the world canvas in his very good history of an Asian trade thoroughfare.

Freedberg, David. *The Eye of the Lynx: Galileo, His Friends, and the Beginnings of Modern Natural History.* Chicago: University of Chicago Press, 2002. The story of Galileo, told in this handsome book, centers on Cesi and the Lynxes, on the drawings of nature brought from the "New" World, and on the crisis in science that came with the introduction of the telescope, the microscope, and the ideas of Galileo.

Gornick, Vivian. *Women in Science: Then and Now.* New York: Feminist Press, 2009. As its title suggests, this is about women pioneers in the scientific world. It describes the time when science was mostly an all-male pursuit and continues to today, when women scientists are winning Nobel Prizes.

Grun, Bernard. *Timetables of History.* New York: Simon & Schuster, 2005. This big compendium is mostly a giant list that brings dates and events together, often citing the unexpected.

Harari, Yuval Noah. *Sapiens: A Brief History of Humankind*. New York: HarperCollins, 2015. This is what the title suggests: a history of humans that is well written and well researched.

Isaacson, Walter. *Leonardo da Vinci*. New York: Simon & Schuster, 2017. This big biography reads like a novel. It provides a portrait of Renaissance Italy along with details on Leonardo's life, his way of work, and his contributions to science.

Jardine, Lisa, and Alan Stewart. *Hostage to Fortune: The Troubled Life of Francis Bacon*. New York: Hill & Wang, 1999. This book tells of political intrigue in Elizabethan England and also explains why Francis Bacon's life has relevance today in the worlds of both political and science history.

Mayr, Ernst. *The Growth of Biological Thought: Diversity, Evolution, and Inheritance*. Cambridge, MA: Belknap, 1985. Mayr begins with Aristotle and then follows scientific ideas into Europe's universities, where subjects such as logic, cosmology, physics, and theology dominate.

McNeill, William H. *The Rise of the West: A History of the Human Community*. Chicago: University of Chicago Press, 1991. This classic of world history is one that I turn to when I want a broad perspective on world events. The 1991 edition includes a retrospective essay.

Morris, Ian. *Why the West Rules—for Now: The Patterns of History, and What They Reveal about the Future*. New York: Farrar, Straus and Giroux, 2010. If I had to pick one world history for you to read, this might be it.

Mukherjee, Siddhartha. *The Gene: An Intimate History*. New York: Scribner, 2016. This is a big but very informative and well-written book that covers the genetic story.

Nuland, Sherwin B. *Doctors: The Biography of Medicine*. New York: Vintage, 1995. Nuland begins his book of stories about doctors with Hippocrates and Galen and Vesalius. He goes on to the twentieth century and the great pediatric physician Helen Taussig.

Oldstone, Michael B. A. *Viruses, Plagues, and History: Past, Present, and Future*. New York: Oxford University Press, 2010. We who have experienced the world-changing coronavirus understand, as many did not in the past, the power of a tiny virus. This is a scholarly book on a tough subject.

Pepys, Samuel. *The Diary of Samuel Pepys*. New York: Everyman's Library, 2018. This is a personal diary of daily life in London from 1660 to 1669, including a pandemic. It's a classic for good reason. It's a wonderful read.

Pomeroy, Sarah B., and Jeyaraney Kathirithamby. *Maria Sibylla Merian: Artist, Scientist, Adventurer*. Los Angeles: J. Paul Getty Museum, 2018. Merian ignored barriers for women in the sciences and arts. Additional resources for readers and teachers are available free online at getty.edu/education/merian.

Ray, Meredith K. *Daughters of Alchemy: Women and Scientific Culture in Early Modern Italy*. Cambridge, MA: Harvard University Press, 2015. This book features women in early modern Europe practicing the healing arts, or alchemy, and making cosmetics; many are midwives. Some, like Margherita Sarrocchi, Camilla Erucliana, and Moderata Fonte, are impressive doers as well as thinkers. And Isabella Cortese? See what you think.

Rutherford, Adam. *A Brief History of Everyone Who Ever Lived: The Human Story Retold Through Our Genes*. New York: Workman, 2016. When Rutherford begins this book, we are naked creatures on the African grasslands. Then he tells a story, packed with information on genes and genomes, of how we came to be who we are today.

Shea, William R., and Mariano Artigas. *Galileo in Rome: The Rise and Fall of a Troublesome Genius*. New York: Oxford University Press, 2003. One of many good books on Galileo, a man who was way ahead of his time.

Snyder, Laura J. *Eye of the Beholder: Johannes Vermeer, Antoni van Leeuwenhoek, and the Reinvention of Seeing*. New York: W. W. Norton, 2015. Two boys of the same age grow up in the same small, vibrant city. Each becomes world-famous: one as an artist, the other as a scientist. Were they friends? This book gives you background so you can decide for yourself.

Wootton, David. *The Invention of Science: A New History of the Scientific Revolution*. New York: HarperCollins, 2015. This serious but readable book strays beyond science into literature and history as well.

Source Notes

Introduction

p. 1: "Courtiers kissed the ground . . . only before God": quoted in Armstrong, 55.

p. 2: "two vast semicircles . . . sides of the river": quoted in Pouwels, 58.

p. 13: "Leonardo was among . . . the blood system": Isaacson, 414.

p. 13: "Leonardo was able . . . supply and nerves": ibid., 415.

Chapter 1

p. 15: "The rise of . . . had in Europe": Armstrong, 141.

p. 15: "The first new science . . . authority of Galen": David Wootton, *The Invention of Science* (New York: HarperCollins, 2015), 302.

p. 19: "A constant element . . . the participant": Erasmus, 114.

p. 19: "When I undertake. . . suit my purpose": Andreas Vesalius, *On the Fabric of the Human Body, Book II: The Ligaments and Muscles*, trans. William F. Richardson and John B. Carman (San Francisco: Norman, 1999), 234.

p. 19: "Galen was deluded . . . but of oxen": Andreas Vesalius, *On the Fabric of the Human Body, Book III: The Veins and the Arteries; Book IV: The Nerves*, trans. William F. Richardson and John B. Carman (San Francisco: Norman, 1998), 114.

p. 19: "Aristotle in particular . . . from the heart": ibid., 160.

p. 19: "Galen never inspected a human uterus": quoted in Annie Bitbol-Hespériès, "Monsters, Nature, and Generation from the Renaissance to the Early Modern Period: The Emergence of Medical Thought," in *The Problem of Animal Generation in Early Modern Philosophy*, ed. Justin E. H. Smith (Cambridge: Cambridge University Press, 2006), 52.

p. 20: "made 240 drawings . . . human body": Isaacson, 400.

p. 21: "The spinal cord . . . to the limbs" and "The muscles which . . . any other animal": ibid., 409.

Chapter 2

p. 29: "The Great Montezuma . . . made of cacao": Bernal Díaz del Castillo, *The True History of the Conquest of New Spain*, trans. Alfred Percival Maudslay (London: Hakluyt Society, 1908).

p. 29: "How can we save . . . and destruction": quoted in Stuart Schwartz, *Victors and Vanquished: Spanish and Nahua Views of the Conquest of Mexico* (Boston: Bedford/St. Martin's, 2000), 212.

p. 32: "pustules that covered people" and "caused great desolation . . . or were blinded": quoted in James Lockhart, ed., *We People Here: Nahuatl Accounts of the Conquest of Mexico* (Berkeley: University of California Press, 1993), 182.

p. 33: "The reason why . . . to enrich themselves": quoted in "Bartolomé de Las Casas," *Encyclopedia Britannica*, last modified June 27, 2021, https://www.britannica.com/biography/Bartolome-de-Las-Casas.

p. 40: "I will be . . . spend my blood": quoted in Benton Rain Patterson, *With the Heart of a King: Elizabeth I of England, Philip II of Spain, and the Fight for a Nation's Soul and Crown* (New York: St. Martin's, 2007), 109.

p. 41: "singeing of the king of Spain's beard": quoted in Bacon, 5: 262.

p. 43: "I know I have the body . . . in the field": "Elizabeth's Tilbury Speech," British Library, http://www.bl.uk/learning/timeline/item102878.html.

p. 43: "And though she be but little, she is fierce": William Shakespeare, *A Midsummer Night's Dream*, ed. Russ McDonald (New York: Penguin, 2016), 55.

Chapter 3

p. 45: "I do not feel obliged . . . attain by them": quoted in Bynum and Porter, 235.

p. 45: "Scientists who become . . . enjoying public acclaim": Dyson, *The Scientist as Rebel*, 282.

p. 47: "Not from the stars . . . or seasons' quality": William Shakespeare, *Shakespeare's Sonnets*, ed. Edward Bliss Reed (New Haven, CT: Yale University Press, 1923), 15.

p. 47: "It is the very error . . . makes men mad": William Shakespeare, *Othello*, ed. Burton Raffel (New Haven, CT: Yale University Press, 2005), 189.

p. 49: "I am as constant as the northern star": William Shakespeare, *Julius Caesar*, ed. Burton Raffel (New Haven, CT: Yale University Press, 2006), 69.

p. 50: "There are more things . . . your philosophy": William Shakespeare, *Hamlet*, ed. Burton Raffel (New Haven, CT: Yale University Press, 2003), 50.

p. 50: "In nature's infinite . . . I can read": William Shakespeare, *Antony and Cleopatra*, ed. Burton Raffel (New Haven, CT: Yale University Press, 2007), 9.

p. 51: "the meeting place . . . minds in Rome": quoted in Ray, 134.

p. 52: "not only acquire . . . things and wisdom" and "display them . . . without any harm": quoted in Mario Livio, *Galileo and the Science Deniers* (New York: Simon & Schuster, 2021), 58.

p. 52: "be slaves neither of Aristotle nor any other philosopher": quoted in Paula Findlen, *Possessing Nature: Museums, Collecting, and Scientific Culture in Early Modern Italy* (Berkeley: University of California Press, 1996), 71.

p. 55: "Most Illustrious and . . . to see all of it": *Galileo's Microscope Anthology* (Florence, Italy: Istituto e Museo di Storia della Scienza), 6, https://brunelleschi.imss.fi.it/esplora/microscopio/dswmedia/risorse/anthology.pdf.

p. 55: "I have contemplated . . . indescribable diligence": ibid.

p.56: "My dear Kepler . . . or shall we cry?": quoted in Frank Durham, *Frame of the Universe: A History of Physical Cosmology* (New York: Columbia University Press, 1983), 144.

p. 59: "necessary evil": quoted in Marguerite Deslauriers, "Lucrezia Marinella," *Stanford Encyclopedia of Philosophy*, 2012, https://plato.stanford.edu/entries/lucrezia-marinella/.

p. 59: "No firmer stability . . . to marvel at": Lucrezia Marinella, *The Nobility and Excellence of Women and the Defects and Vices of Men*, trans. Anne Dunhill (Chicago: University of Chicago Press, 2007), 143.

p. 59: "Some people, possessing . . . the smallest quantity": ibid., 83.

p. 59: "military arts" and "what would men . . . full of admiration?": ibid., 78.

p. 60: "Perhaps a mature . . . for his shoulders": ibid., 121.

p. 61: "The members of that eminent . . . literature and philosophy": *London Exhibited in 1851*, ed. John Weale (London: John Weale, 1851), 538.

p. 61: "recognise, promote, and . . . benefit of humanity": "Mission and Priorities," The Royal Society, https://royalsociety.org/about-us/mission-priorities/.

p. 61: "*Nullius in verba*": "History of the Royal Society," The Royal Society, https://royalsociety.org/about-us/history/.

Chapter 4

p. 63: "They who have presumed . . . philosophy and learning": Bacon, 14: 27.

p. 63: "I profess to learn . . . fabric of Nature.": William Harvey: Lawrence I. Conrad, Michael Nev, Vivian Nutton, Roy Porter, Andrew Wear, *The Western Medical Tradition: 800 BC to AD 1800* (Cambridge UK, Cambridge University Press, 1995), 337.

p. 64: "There happened in my time . . . without loss": quoted in Bacon, 7: xxxviii, note (e).

p. 64: "the wisest, brightest, meanest of mankind": quoted in Catherine Drinker Bowen, *Francis Bacon: The Temper of a Man* (New York: Fordham University Press, 1993), 5.

p. 65: "Revenge is a . . . he is superior": Bacon, 1: 14.

p. 65: "A little philosophy . . . about to religion": ibid., 1: 53.

p. 71: "*Multi pertransibunt et augebitur scientia*": quoted in Brian Vickers, "Francis Bacon and the Progress of Knowledge," *Journal of the History of Ideas* 53, no. 3 (September 1992): 495.

p. 71: "Men have . . . the material world": quoted in George Henry Lewes, *The History of Philosophy: From Thales to Comte* (London: Longmans, Green, 1880), 126.

p. 75: "My very good . . . succeeded excellently well": quoted in John H. Lloyd, *The History, Topography, and Antiquities of Highgate, in the County of Middlesex* (Highgate, UK: Hazell, Watson and Viney, 1888), 239.

p. 75: "Much better is . . . of your days": quoted in *Home Study Circle Library: The World's Greatest Scientists*, ed. Seymour Eaton (New York: Doubleday & McClure, 1900), 38.

p. 75: "He writes philosophy . . . like a lord chancellor": John Rutherfurd Russell, *The History and Heroes of the Art of Medicine* (London: J. Murray, 1861), 198.

Chapter 5

p. 77: "If by some fiat . . . is marine limestone": John McPhee, *Annals of the Former World* (New York: Farrar, Straus and Giroux, 2000), 152.

p. 77: "Should philosophy guide . . . experiments guide philosophy?": Liu Cixin, *The Three-Body Problem* (New York: Tor, 2014), 17.

pp. 79–78: "M. Steno is . . . next to him": quoted in David S. Lux, *Patronage and Royal Science in Seventeenth-Century France: The Académie de Physique in Caen* (Ithaca, NY: Cornell University Press, 1989), 40.

p. 80: "let a whole . . . without difficulty": Troels Kardel and Paul Maquet, *Nicolaus Steno: Biography and Original Papers of a 17th Century Scientist* (Berlin: Springer, 2018), 715.

p. 82: "The stratified stones . . . floods of the rivers": Leonardo da Vinci, *The Notebooks of Leonardo Da Vinci*, vol. 2 (Mineola, NY: Dover, 1960), 980.

Chapter 6

p. 85: "A scientist in . . . a fairy tale": quoted in Eve Curie, *Madame Curie: A Biography by Eve Curie*, trans. Vincent Sheean (New York: Da Capo, 2001), 341.

p. 85: "The senses are scouts . . . to Reason within": Francesco Redi, *Experiments on the Generation of Insects*, trans. Mab Bigelow (Chicago: Open Court, 1909), 19.

pp. 87–88: "Cut two pieces . . . have been generated": quoted in Christopher Wills and Jeffrey Bada, *The Spark of Life: Darwin and the Primeval Soup* (Oxford: Oxford University Press, 2001), 2.

Chapter 7

p. 93: "The next care . . . to the understanding": Hooke, 1: 53.

p. 93: "The abiding fact is . . . ultimate unity of all life": Goldstein, 78.

p. 95: "Physico-Mathematical-Experimental Learning": Henry Lyons, *The Royal Society, 1660–1940* (Cambridge: Cambridge University Press, 1944), 21.

p. 95: "eminent scientists": "History of the Royal Society," The Royal Society, https://royalsociety.org/about-us/history/.

p. 95: "Hooke's book of . . . presently [bought] it": Pepys, https://www.pepysdiary.com/diary/1665/01/02/.

p. 95: "till two o'clock . . . Mr. Hooke's Microscopicall Observations" and "the most ingenious . . . in my life": ibid., https://www.pepysdiary.com/diary/1665/01/21/.

p. 101: "curator of experiments": "Robert Hooke," The Royal Society, https://makingscience.royalsociety.org/s/rs/people/fst00009590.

p. 101: "Knowledge itself is a power": Bacon, 1: 219.

p. 101: “He is of prodigious . . . in the World”: quoted in Robert D. Purrington, *The First Professional Scientist: Robert Hooke and the Royal Society of London* (Boston: Birkhäuser, 2009), 16.

p. 102: “infinity of living animals”: *Galileo's Microscope Anthology* (Florence, Italy: Istituto e Museo di Storia della Scienza, 2018), 10, https://brunelleschi.imss.fi.it/esplora/microscopio/dswmedia/risorse/anthology.pdf.

p. 102: “And now thanks to microscopes . . . tiny bodies”: ibid.

p. 104: “troublesome to be drawn”: Hooke, 203.

p. 104: “knock'd him down dead drunk . . . moveless,” “suddenly reviv'd and ran away,” and “several bubbles”: ibid., 204.

p. 104: “a great price . . . in the world”: Pepys, https://www.pepysdiary.com/diary/1664/08/13/.

p. 104: “with great difficulty”: ibid., https://www.pepysdiary.com/diary/1664/08/14/.

Chapter 8

p. 107: “Our universe is . . . and for all”: Seneca, *Natural Questions,* vol. 1, trans. Thomas H. Corcoran (Cambridge, MA: Harvard University Press, 1989), 293.

p. 107: “We may hope . . . all of science”: Dyson, *The Scientist as Rebel*, 185.

p. 112: “I am writing . . . of his work”: quoted in Dobell, 40–41.

p. 112: “Our honest citizen . . . as this man is”: ibid., 43.

p. 113: “The globules in . . . black taffety silk”: ibid., 331.

p. 113: “I had no thought . . . drop of water”: ibid., 123.

p. 113: “Observing again, I . . . in one drop”: ibid., 124.

p. 113: “swam gently among . . . in the air” and “had a much swifter . . . a quick dart”: ibid.

p. 113: “so small, in . . . of coarse sand”: ibid., 133.

p. 113: “a little white . . . if 'twere batter”: ibid., 239.

p. 116: “In Holland I marveled . . . East and West Indies” and “This prompted me . . . expensive journey”: quoted in E. Charles Nelson and David John Elliott, eds., *The Curious Mister Catesby* (Athens: University of Georgia Press, 2015), 46.

p. 117: “The work consists of . . . rare toads and frogs”: Kay Etheridge, “The History and Influence of Maria Sibylla Merian's Bird-Eating Tarantula: Circulating Images and the Production of Natural Knowledge,” in *Global Scientific Practice in the Age of Revolutions*, 1750–1850, ed. Patrick Manning and Daniel Rood (Pittsburgh: University of Pittsburgh Press, 2016), 58.

p. 118: “an unbelievably great . . . in going forwards”: quoted in Dobell, 241.

p. 118: "seemed to be alive": ibid., 242.

p. 118: "to go over to the Truth, and to cleave unto it": ibid., 74.

p. 118: "My work, which . . . be informed thereof": ibid., 82–83.

Chapter 9

p. 121: "What's in a name? . . . smell as sweet": William Shakespeare, *Romeo and Juliet*, ed. Burton Raffel (New Haven, CT: Yale University Press, 2004), 60.

p. 121: "I pull a flower . . . in a 'class'!": Emily Dickinson, *The Poems of Emily Dickinson*, ed. Thomas H. Johnson (Cambridge, MA: Belknap/Harvard University Press, 1955), 56.

p. 122: "was in reality . . . become a naturalist": quoted in Albert Martini, *The Renaissance of Science: The Story of the Cell and Biology* (Maitland, FL: Abbott Communications Group, 2015), 355.

p. 123: "For riches vanish . . . aloft in botany": quoted in Nelis A. van Der Cingel, *An Atlas of Orchid Pollination: European Orchids* (Brookfield, VT: A. A. Balkema, 2001), 38.

p. 125: *God Have Mercy on This House*: quoted in Stephen Coss, *The Fever of 1721: The Epidemic That Revolutionized Medicine and American Politics* (New York: Simon & Schuster, 2016), 72.

p. 125: "had undergone an operation . . . preserve him from it" and "he described the operation . . . had left upon him": quoted in George Lyman Kittredge, *Some Lost Works of Cotton Mather* (Cambridge, MA: John Wilson and Son, 1912), 422.

p. 127: "Names have the same value . . . in public affairs": quoted in Jan Sapp, *The New Foundations of Evolution: On the Tree of Life* (New York: Oxford University Press, 2009), 6.

p. 127: "We are very fond . . . books in England": quoted in P. J. Marshall, ed., *The Oxford History of the British Empire*, vol. 2: *The Eighteenth Century* (New York: Oxford University Press, 2006), 239.

p. 128: "'homo' is a . . . to all men": William Shakespeare, *Henry IV, Part I*, ed. Claire McEachern (New York: Penguin, 2017), 30.

p. 128: "I know no greater man on earth" and "*Deus creavit, Linnaeus disposuit*": quoted in Kennedy Warne, "Organization Man," *Smithsonian*, May 2007, https://www.smithsonianmag.com/science-nature/organization-man-151908042/.

p. 128: "the Prince of botanists": quoted in Richard Collins, "Celebrating the 'Prince of Botanists,'" *Irish Examiner*, November 13, 2006, https://www.irishexaminer.com/opinion/columnists/arid-20018103.html.

p. 128: "With the exception . . . me more strongly": quoted in Patricia Hampl, *The Art of the Wasted Day* (New York: Penguin, 2019), 193.

Chapter 10

p. 133: "I shall collect . . . harmony in nature": quoted in Laura Dassow Walls, *Seeing New Worlds: Henry David Thoreau and Nineteenth-Century Natural Science* (Madison: University of Wisconsin Press, 1995), 96.

p. 133: "I write this . . . of the world": Thomas Jefferson to Alexander von Humboldt, June 13, 1817, National Archives, https://founders.archives.gov/documents/Jefferson/03-11-02-0361.

p. 134: "one of those . . . world, like Aristotle": Ralph Waldo Emerson, *The Complete Works of Ralph Waldo Emerson*, vol. 11 (Boston: Houghton, Mifflin, 1904), 457.

p. 134: "compare the observations . . . of miles apart": Andrea Wulf, *The Invention of Nature: Alexander Von Humboldt's New World* (New York: Vintage, 2016), 4.

p. 136: "reaches further back . . . hitherto been suspected": quoted in Richard Owen, *Paleontology* (Edinburgh: Adam and Charles Black, 1861), 18.

p. 139: "*omnis cellula e cellula*": "Rudolf Virchow," in Oxford Reference, https://www.oxfordreference.com/view/10.1093/oi/authority.20110803115939259.

p. 140: "who expects to . . . without the telescope": quoted in John Phin, *Practical Hints on the Selection and Use of the Microscope* (New York: Industrial Publication Company, 1877), 11.

pp. 141, 143: "are very inferior . . . all the animals": quoted in Jan Sapp, *The New Foundations of Evolution: On the Tree of Life* (New York: Oxford University Press, 2009), 18.

p. 143: "rare union of poetry with science": quoted in *Scientific Discovery: Case Studies*, ed. Thomas Nickles (Boston: D. Reidel, 1980), 117.

Chapter 11

p. 145: "Bacteria represent the . . . on this planet": Stephen Jay Gould, "The Evolution of Life on the Earth," in *Evolution: A Scientific American Reader* (Chicago: University of Chicago Press, 2006), 240.

p. 145: "We've tried ignorance . . . we try education": Joycelyn Elders, "Three-Ring Circus," *Advocate*, October 31, 1995, 60.

p. 150: "Honored Professor! . . . of *Bacillus anthracis*": quoted in Richard Adler, *Robert Koch and American Bacteriology* (Jefferson, NC: McFarland, 2016), 22.

Chapter 12

p. 157: "All that we can say . . . into a cell": Dyson, *Origins of Life*, 28.

p. 157: "There is a tide . . . we now afloat": William Shakespeare, *Julius Caesar*, ed. William Montgomery (New York: Penguin, 2016), 92.

p. 161: "is bacterial, but . . . yet been isolated": quoted in Arnold J. Levine and Lynn W. Enquist, "History of Virology," in *Fields Virology*, ed. David M. Knipe and Peter M. Howley (Philadelphia: Lippincott Williams & Wilkins, 2007), 4.

Looking Ahead

p. 165: "A man who dares . . . value of life": Charles Darwin, *The Autobiography of Charles Darwin* (New York: Barnes and Noble, 2005), 156.

Bibliography

Arikha, Noga. *Passions and Tempers: A History of the Humours*. New York: HarperCollins, 2007.

Armstrong, Karen. *Islam: A Short History*. New York: Modern Library, 2000.

Bacon, Francis. *The Works of Francis Bacon, Lord Chancellor of England*. Edited by Basil Montagu. London: William Pickering, 1827. https://catalog.hathitrust.org/Record/001915361/Home.

Bryson, Bill. *The Body: A Guide for Occupants*. New York: Doubleday, 2019.

———. *Seeing Further: The Story of Science, Discovery, and the Genius of the Royal Society*. New York: HarperCollins, 2010.

Bynum, W. F., and Roy Porter. *Oxford Dictionary of Scientific Quotations*. New York: Oxford University Press, 2005.

Calder, Nigel. *Magic Universe: The Oxford Guide to Modern Science*. Oxford: Oxford University Press, 2003.

Carroll, Sean B. *The Big Picture: On the Origins of Life, Meaning, and the Universe Itself*. New York: Dutton, 2017.

———. *The Serengeti Rules: The Quest to Discover How Life Works and Why It Matters*. Princeton, NJ: Princeton University Press, 2016.

———. *The Story of Life: Great Discoveries in Biology*. New York: W. W. Norton, 2019.

Challoner, Jack. *The Cell: A Visual Tour of the Building Block of Life*. Chicago: University of Chicago Press, 2017.

Dobell, Clifford. *Antony van Leeuwenhoek and His "Little Animals": Being Some Account of the Father of Protozoology and Bacteriology and His Multifarious Discoveries in These Disciplines*. New York: Dover, 1960.

Dyson, Freeman. *Disturbing the Universe*. New York: Basic Books, 1979.

———. *Origins of Life*. Cambridge: Cambridge University Press, 2004.

———. *The Scientist as Rebel*. New York: New York Review of Books, 2006.

Erasmus, Desiderius. *Collected Works of Erasmus: Correspondence*. Edited by Richard J. Schoeck and Beatrice Corrigan. Toronto: University of Toronto Press, 1974.

Farrell, Jeanette. *Invisible Enemies: Stories of Infectious Diseases*. New York: Farrar, Straus and Giroux, 2005.

Freedberg, David. *The Eye of the Lynx: Galileo, His Friends, and the Beginnings of Modern Natural History*. Chicago: University of Chicago Press, 2002.

Godfrey-Smith, Peter. *Metazoa: Animal Life and the Birth of the Mind*. New York: Farrar, Straus and Giroux, 2020.

Goldstein, Thomas. *Dawn of Modern Science*. Boston: Houghton Mifflin, 1980.

Gornick, Vivian. *Women in Science: Then and Now*. New York: Feminist Press, 2009.

Grun, Bernard. *Timetables of History*. New York: Simon & Schuster, 2005.

Harari, Yuval Noah. *Sapiens: A Brief History of Humankind*. New York: Vintage, 2011.

Harold, Franklin M. *The Way of the Cell: Molecules, Organisms and the Order of Life*. New York: Oxford University Press, 2001.

Hooke, Robert. *Micrographia*. London: Jo. Martyn and Ja. Allestry, 1665. https://www.loc.gov/item/11004270/; https://www.gutenberg.org/files/15491/15491-h/15491-h.htm.

Isaacson, Walter. *Leonardo da Vinci*. New York: Simon & Schuster, 2017.

Jardine, Lisa, and Alan Stewart. *Hostage to Fortune: The Troubled Life of Francis Bacon*. New York: Hill & Wang, 1999.

Lev, Elizabeth. *The Tigress of Forli: Renaissance Italy's Most Courageous and Notorious Countess, Caterina Riario Sforza de' Medici*. New York: Houghton Mifflin Harcourt/Mariner, 2011.

Lightman, Alan. *The Discoveries: Great Breakthroughs in 20th-Century Science*. New York: Vintage, 2006.

Mayr, Ernst. *The Growth of Biological Thought: Diversity, Evolution and Inheritance*. Cambridge, MA: Belknap, 1985.

McNeill, J. R., and William McNeill. *The Human Web: A Bird's-Eye View of World History*. New York: W. W. Norton, 2003.

McNeill, William H. *The Rise of the West: A History of the Human Community*. Chicago: University of Chicago Press, 1963.

Milner, Richard. *Darwin's Universe: Evolution from A to Z*. Oakland: University of California Press, 2009.

Morris, Ian. *Why the West Rules—for Now: The Patterns of History, and What They Reveal about the Future*. New York: Farrar, Straus and Giroux, 2010.

Moulton, F. R., and Justus J. Schiffere. *The Autobiography of Science*. New York: Doubleday, 1960.

Mukherjee, Siddhartha. *The Gene: An Intimate History*. New York: Scribner, 2016.

Nuland, Sherwin B. *Doctors: The Biography of Medicine*. New York: Vintage, 1995.

Nurse, Paul. *What Is Life? Five Great Ideas in Biology*. New York: W. W. Norton, 2021.

Oldstone, Michael B. A. *Viruses, Plagues, and History: Past, Present, and Future*. New York: Oxford University Press, 2010.

Pepys, Samuel. *The Diary of Samuel Pepys*. Edited by Henry B. Wheatley. London: George Bell and Sons, 1893. The Diary of Samuel Pepys, https://www.pepysdiary.com/.

Pomeroy, Sarah B., and Jeyaraney Kathirithamby. *Maria Sibylla Merian: Artist, Scientist, Adventurer*. Los Angeles: J. Paul Getty Museum, 2018.

Pouwels, Randall L. *The African and Middle Eastern World, 600–1500*. New York: Oxford University Press, 2005.

Ray, Meredith K. *Daughters of Alchemy: Women and Scientific Culture in Early Modern Italy*. Cambridge, MA: Harvard University Press, 2015.

Reich, David. *Who We Are and How We Got Here: Ancient DNA and the New Science of the Human Past*. New York: Pantheon, 2018.

Rutherford, Adam. *A Brief History of Everyone Who Ever Lived: The Human Story Retold Through Our Genes*. New York: Workman, 2016.

Shea, William R., and Mariano Artigas. *Galileo in Rome: The Rise and Fall of a Troublesome Genius*. New York: Oxford University Press, 2003.

Snyder, Laura J. *Eye of the Beholder: Johannes Vermeer, Antoni van Leeuwenhoek, and the Reinvention of Seeing*. New York: W. W. Norton, 2015.

Zimmer, Carl. *Life's Edge: The Search for What It Means to Be Alive*. New York: Dutton, 2021.

———. *She Has Her Mother's Laugh: The Powers, Perversions, and Potential of Heredity*. New York: Dutton, 2018.

Image Credits

Biodiversity Heritage Library: pp. 69 (bottom right), 137, 140, 160 (bottom)

Library of Congress, Geography and Map Division: pp. viii–1, 10–11, 43, 76–77, 131 (top), 158

Library of Congress, Prints & Photographs Division: pp. 21, 34 (bottom), 39 (top), 44–45, 46 (top), 54, 64, 69 (top), 71, 73 (bottom), 82, 84–85, 95 (bottom), 98 (left), 124 (both), 128, 131 (bottom), 139 (top), 143, 144–145, 153

Library of Congress, Rare Book & Special Collections Division: front and back cover (top right and bottom left), pp. i, ii, 9 (both), 26 (top), 41 (both), 47 (both), 48 (all), 52, 66 (top), 67 (both bottom images), 78, 79, 88, 89 (both), 94 (left), 110 (middle), 112, 114, 120–121, 127, 135 (top), 154

Medicea Laurenziana Library, Florence/World Digital Library: pp. 30 (bottom), 32, 37 (both)

Courtesy of the National Library of Medicine: back cover (center), pp. 14–15, 16, 17 (both), 18, 23 (both), 25, 26 (bottom), 66, 67 (top), 74, 86, 87, 90 (top), 91, 94 (both right images), 96 (both), 98 (right), 99 (bottom), 103, 104, 108, 125, 126, 147 (bottom), 149 (top), 150, 151 (both), 152, 159

Courtesy of the Science History Institute: pp. 106–107, 109, 111, 155 (top)

Front and back cover: The Walters Museum (top left); Vue du Cajambe, Wellcome Collection, Attribution 4.0 International (CC BY 4.0) (bottom right)

Front cover (center): Courtesy of Biblioteca Apostolica Vaticana, Città del Vaticano

pp. 1 (both right images) and 55: Museo Galileo, Firenze; photo by Franca Principe

p. 2 (top): Jean Soutif/Look at Sciences/Science Photo Library

p. 2 (bottom): bpk Bildagentur/Vorderasiatisches Museum, Berlin, Germany/Olaf M. Tessmer/Art Resource, NY

p. 3: Facsimile by Hugh R. Hopgood. Rogers Fund 1930, Metropolitan Museum of Art

p. 4: The Miriam and Ira D. Wallach Division of Art, Prints and Photographs: Picture Collection, the New York Public Library

p. 6: © The National Gallery, London

p. 7: The Walters Art Museum

p. 8: Jefferson Building, Library of Congress

pp. 12 and 20 (bottom): Royal Collection Trust/© Her Majesty Queen Elizabeth II 2022

p. 20 (top): Vesalius stamp designed by János Kass used with permission from the Philatelic Department, Hungarian Postal Service

p. 22: Photo by dchelyadnik@yahoo.com/Flickr

p. 24: The Picture Art Collection/Alamy Stock Photo

pp. 28–29: Tula, Pyramid B, Atlantes/Photo by Arian Zwegers/Flickr/CC BY 2.0

p. 30 (both top images): Library of Congress, Jay I. Kislak Collection

p. 31: The Michael C. Rockefeller Memorial Collection at the MET, Gift of Nelson A. Rockefeller, 1969

p. 33: Internet Archive/John Carter Brown Library

p. 34 (top): Digital image courtesy of the Getty Museum's Open Content Program

p. 35 (top): Pixabay

p. 35 (bottom): Photo by Alexandre Buisse, CC BY-2.0, Wikimedia Commons

p. 36: © Photographic Archive, Museo Nacional del Prado

p. 38 (both): Library of the National Academy of the Lincei and of the Corsini Family/World Digital Library

p. 39 (bottom): © PromoMadrid, author Max Alexander

p. 42: IanDagnall Computing/Alamy Stock Photo

p. 46 (bottom): Photograph: CliveSherlock.com

p. 49: National Library of Scotland

p. 50: By permission of the Accademia Nazionale dei Lincei, Rome

p. 51 (top): National Central Library of Florence/World Digital Library

p. 51 (bottom): Beinecke Rare Book and Manuscript Library, Yale University

p. 53: Courtesy of Biblioteca Apostolica Vaticana, Città del Vaticano

p. 56: Zentralbibliothek Zürich

p. 58: Wiki Commons

p. 59: Biblioteca Correr, Venezia/Musei Civici di Venezia

p. 60: Courtesy of the Folger Shakespeare Library

pp. 62–63: © National Portrait Gallery, London

p. 65: Internet Archive/Duke University Libraries

p. 69 (bottom left): Courtesy of Special Collections, Toronto Public Library

p. 70: Caroline Gluck/Flickr/Oxfam CC BY 2.0

p. 73 (top): Universal Images Group North America LLC/DeAgostini/Alamy Stock Photo

pp. 80 and 81 (both): Internet Archive/University of Maryland/Baltimore Digital Archive

p. 90 (bottom): Tripadvisor/Dorian C.

pp. 92–93: Courtesy of the Wellcome Collection

p. 95 (top): Complutense University of Madrid/World Digital Library

p. 99 (top): GC6.Sch256.630r, Houghton Library, Harvard University

p. 100: Harris Brisbane Dick Fund, 1953/Metropolitan Museum of Art

p. 101: Courtesy of the National Museum of Health and Medicine

p. 105 (left): John Montenieri, Centers for Disease Control and Prevention

p. 105 (right): Llyfgrell Genedlaethol Cymru

p. 110 (top): Städel Museum, Frankfurt am Main

p. 110 (bottom): Division of Medicine and Science, National Museum of American History, Smithsonian Institution

p. 113: Ann Ronan Picture Library/Heritage-Images

p. 115: The Albertina Museum, Vienna

p. 116: The British Library

p. 117: Internet Archive/Getty Research Institute

p. 119: Wellcome Collection, Attribution 4.0 International (CC BY 4.0)

p. 123 (both): Slovak National Library/World Digital Library

p. 132–133: Vue du Cajambe, Wellcome Collection, Attribution 4.0 International (CC BY 4.0)

p. 134: bpk Bildagentur/Staatliche Museen, Berlin/Photo by Klaus Goeken/Art Resource, NY

pp. 135 (bottom) and 142: Museum of Natural History, Berlin

p. 136: Internet Archive

p. 138 (top): Photo by Nathan Shankar, University of Oklahoma; digital colorization by Stephen Ausmus

p. 138 (bottom): Alison R. Taylor (University of North Carolina Wilmington Microscopy Facility)

p. 139 (bottom): Hathi Trust

p. 141: BCCB/Photographer: Fayette A Reynolds, MS

p. 146: National Institute of Allergy and Infectious Diseases

p. 147 (top): University of Wisconsin/Hathi Trust

p. 149 (bottom): Internet Archive/American Libraries

p. 155 (bottom): Institut Pasteur/photo by François Gardy

pp. 156–157 and 160 (top): With permission of the Curator, the Delft School of Microbiology Archives, Department of Biotechnology, Delft University of Technology, the Netherlands

p. 161: heidICON, University Library Heidelberg

p. 162 (top): Copyright © The Royal Society

p. 162 (middle and bottom): Steve Gschmeissner/Science Photo Library

p. 165: Wellcome Library/World Digital Library

Index

Page numbers in *italics* indicate images and/or captions. Pages numbered in roman type may also contain relevant images and/or captions.

JOY HAKIM is the best-selling author of A History of US, a ten-volume history of the United States, which has sold more than three million copies, as well as of the much-lauded The Story of Science series. She has worked as a teacher, journalist, and editor and lives in Maryland.